Put Your Science to Work

The Take-Charge Career Guide for Scientists

Peter S. Fiske

Illustrated by Aaron Louie

AGU
Washington, DC
2001

Published under the aegis of the AGU Education and Human Resources Committee

Library of Congress Cataloging-in-Publication Data

Fiske, Peter S., 1966-
Put Your science to work: the take-charge career guide for scientists/Peter S. Fiske
p. cm.
Rev. ed. of: To Boldly Go. 1996
Includes bibliographical references and index.
ISBN 0-87590-295-2
1. Science--Vocational guidance.
I. Fiske, Peter S., 1966- To Boldly Go. II. Title.
Q147.F58 2000
502'.3--dc21 00-052578
CIP

The transferable skills (page 12) and personal qualities (page 13) lists are modified, with permission, from lists compiled by Stanford Career Planning and Placement Center, Stanford University, California.

American Geophysical Union
2000 Florida Avenue, N.W.
Washington, DC 20009

Printed in the United States of America

Contents

Foreword

The training of graduate scientists and engineers is a crucial investment—one that provides great dividends by producing both the knowledge and the personnel that America needs, if we are to remain a leading nation in the twenty-first century.

The job market today presents challenges and opportunities for young scientists. Many new graduates remain concerned about the "traditional" job market in academic research and teaching. Yet, from my viewpoint in Washington, DC, it is clear to me that we also need scientifically educated people in many other places besides universities. The traditional value system in academia has seemed to be: you are not really a scientist unless you are actively doing research. Research has been the litmus test, regardless of its quality.

This attitude is changing today. When I took my first job at Princeton University as an assistant professor in the Department of Chemistry, I unconsciously adopted the attitude of other professors. I thought that my job was to take these bright young undergraduates, decide who could really do science like mine and who couldn't, and get those young people who were not like me in their interests or abilities out of science and into some other university department.

In order to get our faculty members excited about the true breadth of career opportunities out there for the next generation of scientists, we need to get them to adopt an enlarged view of who is a scientist. We must expand our conventional view of the scientific community and invite all those scientists who have turned journalist, teacher, policy maker, or whatever back to our science departments on our campuses to tell their stories and to act as role models and mentors for students.

We older scientists have an obligation to younger scientists: we must offer you a broader pathway for using your science in productive careers. Our nation needs many more scientifically trained young people. But the scientific community must broaden its view of who is a scientist, and what constitutes a successful career for someone with a strong scientific education.

I hope that this book helps you recognize your strengths, identify your opportunities, and explore your options.

I wish you the best of luck in your future endeavors.

Dr. Bruce Alberts
President, National Academy of Sciences

1996
Shrinking Job Market
Young Physicists Hear Wall Street Calling
Job Market Blues
WILL WORK FOR FOOD.
No Ph.D.s Need Apply
Science: The government said
Congress Holds Hearings On S&E Workfor
Proposes Maj
Increase In
Science Fundi
Lowest in 25 Years
Internet
R&D
2001
Biotech

Preface

There is a tendency on the part of faculty to want to clone themselves and, by their attitude, to make students feel that "success" means a career in research at a university or at one of the few large industrial laboratories that are left. This tendency is misguided, for most jobs for our graduates have always been in industry and not in research. One of the reasons society supports us is to train people who will transform the work done at universities into something of more direct benefit to society.

Burton Richter, 1995
Past President, American Physical Society

What a Difference 5 Years Makes!

In 1996, when the first edition of this book (*To Boldly Go*) hit the bookstands, it was a grim time for young scientists. Across all fields of science, newly minted Ph.D.s and Masters students were facing the combined effects of falling employment for young scientists, rising Masters and Ph.D. production, and a glut of job seekers in academia. The crisis was covered in leading newspapers and news magazines and was even featured in the popular cartoon series "Doonesbury."

Fast-Forward 5 Years.

The landscape of science employment has changed dramatically. Unemployment rates in the United States have fallen to historically low levels and economic growth and low inflation have fueled one of the longest economic expansions in American history. Much of the growth in the "New Economy" has been stimulated by the innovations of scientists and engineers. Some fields in science and technology are so hot that graduate departments are scarcely able to keep their students in their seats. Given today's booming economy and low unemployment figures, especially in the technology sector, it might be tempting to conclude that the scientist glut of the early 1990s was an aberration, a temporary downturn on an otherwise robust path of growth.

Today, young scientists face a dizzying array of career choices that were barely conceivable 5 years ago. Entire new scientific disciplines have sprung up to address new opportunities in biology, engineering, mathematics, and computer science. Universities are becoming increasingly entrepreneurial, cultivating partnerships with technology companies and building start-up business incubators. Twenty-two-year-old computer science grads are starting their own companies. These would seem to be the best of times for a young, smart person such as yourself.

However, amidst all this innovation and growth, graduate education in the sciences hasn't changed very much. Despite calls for change from the National Research Council, the U.S. Congress, professional societies, and many individuals, graduate education in the United States still focuses on the preparation of young scholars for careers in academia, a minority employer of today's Ph.D. scientists and engineers.

Young scientists today are asking a range of questions about career prospects and opportunities that their advisors, department chairs, and universities are unable to answer. This process of exploration is often difficult and frustrating. While in graduate school, students are rarely exposed to career fields outside research science and, at its root, graduate education remains a process of apprenticeship in which students prepare themselves for a life in science. Having completed an advanced degree, many graduates find themselves far from their schools, without access to on-campus career centers and other resources that can provide information and counseling.

To be fair, graduate students and young scientists are as much to blame for our current job predicament as the institutions that trained us. Very few of us objectively surveyed the landscape of the research science career, weighed the relative merits and drawbacks of the lifestyle, or dispassionately asked ourselves if the geometric growth that employed our advisors could continue indefinitely. Most of us went to graduate school because we loved doing science, we were good at it, and at the time it seemed a relatively secure profession. We pitied our college friends who spent their senior years applying for job after job, and we assumed that the hard time we would spend in graduate school would allow us to side-step such unpleasantness. In reality, we simply deferred it for a while.

This Book Is About Creating Options and Recognizing Opportunities

Career planning is a process of professional development that is important for every type of career, including research science. This book is not an exhortation for you to abandon your research career goals. Rather, its goal is to show you that a wealth of opportunities exist for you in many career fields, especially because you have an advanced degree in science. Far from being a liability, a scientific training provides powerful problem-solving tools that are valuable in nearly every type of career. We scientists have much to offer the world beyond scholarly research. Ph.D. and Masters degree holders do encounter perceptions from the scientific community, the "outside world," and even within themselves that tend to reduce their career options. This book will help you attack those preconceptions and explore your true range of career options.

Exploring alternative careers can be a liberating, empowering, and enjoyable experience. Who knows? Maybe your exploration will confirm your original career goals. No matter what the outcome, you will be better off for the experience both in terms of your own career development and in the advice you may give to your students in the future.

Only you can be in control of your career and nobody cares more than YOU about your future.

Acknowledgments

This second edition would not have been possible without the support, encouragement, ideas and suggestions of many individuals. I am grateful to those who made direct contributions to this edition. I thank all the individuals profiled in Chapter 2 for allowing their interesting stories to be retold, in some cases, for a second time. I am also grateful to the individuals whose true stories and resumes form the basis for the case studies in Chapter 10. I am indebted to my colleague and friend Margaret Newhouse for allowing the use of some of her self-assessment exercises in Chapter 5, and to Stanford University's Career Planning Center for use of their career planning pyramid. Al Levin guided me through the field of career counseling for the first edition. His guidance resonates through this edition as well. Chris Gales edited an early version of this edition. Finally, I am grateful to my cartoonist, Aaron Louie, for wielding his wit and pen on this project.

I have benefited enormously from the suggestions, ideas, and support of young scientists, activists, administrators, and career counselors around the country. Wendy Yee and Nicole Ruediger, two founding staff members of *Science's* Next Wave (www.nextwave.org), gave early support and encouragement for this second edition and have provided me with information and resources along the way. Emily Klotz, Crispin Taylor, and the current staff of Next Wave have continued that support and encouragement. Members of the Association for Science Professionals (ASP), especially Victoria McGovern, Steve Smith, Kevin Aylesworth, Finley Austin, Bob Rich, and Patricia Bresnahan, have fed me stories, statistics and research material that have greatly enriched this edition. Geoff Davis, my friend and partner on the Grad School Survey (http://survey.nagps.org/frontpage.shtml), provided a wealth of material through his web site, Phds.org. Eliene Augenbraun and Kelly Kirkpatrick supplied ground truth and much more.

Many of the changes in this edition have come about from conversations I have had with the students, post-docs, administrators, career counselors, and the parents of young scientists I have met in the course of my lectures around the country. Thank you all for sharing your thoughts and insights.

My heartfelt thanks go to Michael Teitelbaum and the Alfred P. Sloan Foundation for supporting this project through a grant to the Center for Science and the Media, and to Lawrence Livermore National Laboratory for allowing me leave to pursue this endeavor.

Finally, this book would not have the quality, depth, and accuracy it has without the care and editorial precision of Jennifer Giesler. Her tireless efforts to improve and extend career development services at AGU, and her research on the job market, have improved the lives and careers of many young scientists. Her career data for geoscientists can be found at: http://www.agu.org/sci_soc/cpst/employment_survey.html.

Science has a great tradition of unselfish cooperation and community service. In this spirit I hope you, the reader, will share your thoughts, observations, and suggestions about jobs, careers, and this book with me and other readers. AGU has set up a companion web site for this book which will feature additional information and resources. Please visit us at: http://www.agu.org/careerguide

Peter S. Fiske, September 14, 2000

peterfiske@yahoo.com

"Let's see. You might get a job. . . or maybe not. Hang on . . . nope. Wait. Yes, you will. I think . . ."

Beyond the Event Horizon

Science Employment Trends in the New Millennium

1

The size of the scientific enterprise, which began its expansion around 1700, has now begun to reach the limits imposed on it by the size of the human race.

David Goodstein
Scientific Elites and Scientific Illiterates
1993 Sigma Xi Forum

A decade ago, Richard Atkinson, then incoming president of the American Association for the Advancement of Science (AAAS), declared the supply of scientists and engineers in the United States a "national crisis in the making." Atkinson was responding to projections by the National Science Foundation (NSF) of a looming shortage of new scientists. Subsequent investigation by young scientists and Congressional staffers revealed that the NSF's projections were dead wrong. As a result, 5 years of science and engineering graduate students marched optimistically into one of the worst job markets for scientists in the past 40 years.

Predicting supply and demand in employment has always been a perilous activity. While near-term supply is fairly easy to judge given the number of students in the pipeline, estimating demand for newly trained scientists and engineers is a black art at best! Not only is it difficult to estimate future hiring trends in academia, industry, and government, but these estimates are predicated on economic and federal policy conditions that can change dramatically over the time frame of a single graduate student. Put simply, there is no way for entering graduate students to know what the job market will be like when they graduate.

This does not mean that graduate students must march into a black hole of uncertainty. While job supply may be difficult to gauge, it is possible to understand some of the macroscopic forces that affect science employment and think strategically about your career in science by asking some basic questions.

What will federal R&D support look like in my future?

As you know, money is the mother's milk of science. Practically every measure of growth in science (number of Ph.D.s, number of publications) is highly correlated to the amount of research funding provided by the government. Most of the R&D money spent in the United States is spent by industry. Basic research—the land most academic Ph.D.s and researchers inhabit—represents only 15% of the total amount of R&D spending, but the federal government funds most of this.

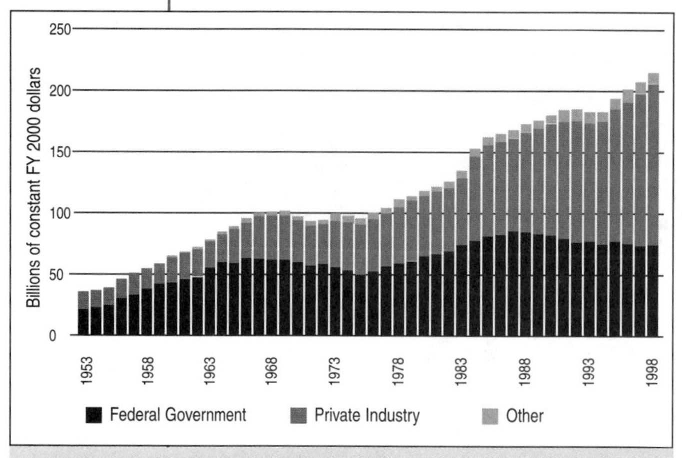

Figure 1. Growth in R&D spending over the last 50 years. The private sector now accounts for two of every three dollars invested in R&D.

The AAAS has amassed figures and data on trends in R&D funding in the United States and abroad. The data, summarized in Figure 1, show that the federal government's spending in R&D has been more or less constant over the past 25 years. Nearly all the growth in total R&D spending has been in the industrial sector. This investment ebbs and flows depending on the health of the economy and the health of particular sectors. "Rich" sectors, such as information technology and biotechnology, spend proportionately more on research while sectors that are highly competitive and have low profit margins, such as the steel industry, tend to shave their R&D investments.

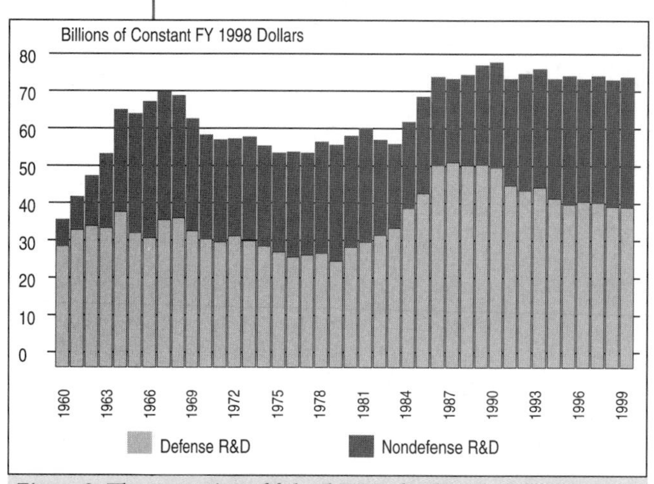

Figure 2. The proportion of federal R&D funding to defense and non-defense areas has changed. Today, nearly half goes to non-defense R&D.

So, in answer to your question, government spending on science and technology probably won't grow much faster than the rate of inflation during your career. Most of the growth will be in industry.

Are there important trends in how the U.S. government is investing in science and technology?

Indeed there are. The proportion of defense-related R&D has fallen substantially since 1990 (see Figure 2). This is due not only to the end of the Cold War but also to an increasing reliance on "Commercial Off The Shelf" (COTS) technology in new defense systems. The Defense Department has substantially cut its basic science funding, relying on industry to come up with the innovations it will need for the future. The needs of the military are changing as well. The United States no longer faces a single, large, technologically comparable adversary. Today we live in a world of numerous small threats that include terrorist groups not aligned with any particular country. As a result, Defense R&D will likely continue to shift toward information technology; light, mobile, and precise weapons systems; and defense against weapons of mass destruction.

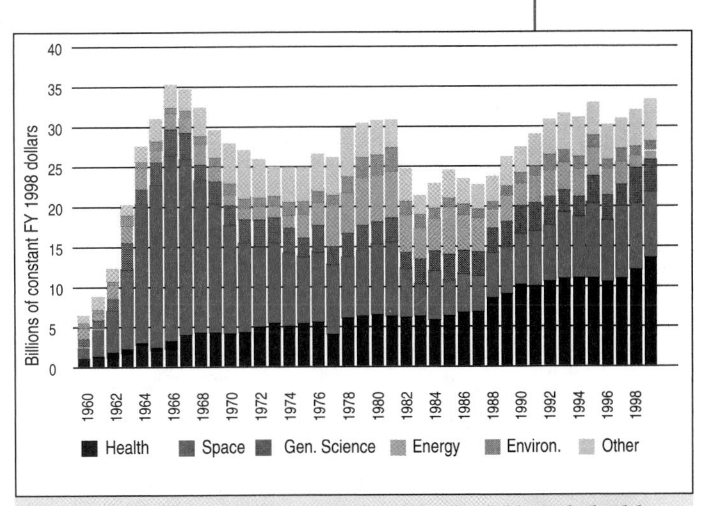

Figure 3. Trends in Federal R&D spending by field. Only health-related research has grown steadily.

In the non-defense part of federal R&D spending only one field of research has shown steady increases in funding year in and year out: Health. Other areas, such as space and energy, have waxed and waned as priorities shifted and crises passed. With the graying of the American electorate, and the huge direct costs borne by the government for health care, one can only expect the proportion of R&D funding for health to increase in the future.

So, in answer to your question, the life and health sciences appear to have the rosiest futures for federal funding. But, as we discussed earlier, the overall level of federal funding for science will not rise dramatically during your career.

But what about those calls in Congress for doubling of science funding?

It is true that in the last few years several members of Congress have called for a "doubling" of funding for science over the next 5 years. Science seems to enjoy popular bipartisan support these days, and many members of Congress believe that the federal investment in science has substantial economic rewards. However, before any of you young scientists get your hopes up, let me caution you that we have heard these words before. Congressional calls for more funding are just that: recommendations. But when it comes to slicing up the shrinking wedge of "discretionary spending"—those federal dollars that are not already committed to Social Security and other entitlement programs—science has to compete with all those other hungry mouths: education, transportation, housing, etc. While science may pay out big dividends in the long term, other "investments" pay far more handsomely in the short term, a.k.a. the Congressional term! Until R&D can compete better with these short-term issues it is likely that federal R&D spending will not get a substantially larger slice of the pie.

Will industrial R&D continue to grow?

Industrial funding of R&D as a whole has been growing for some time, and there is every indication that, as the economy grows, industrial R&D will grow as well. Some new industries, such as information technology and biotechnology, are R&D-intensive. As these grow they will draw in more scientific talent. Some in industry and government are worried that rapid growth in these industries will be limited not by funding but by a lack of technical professionals to fill new jobs.

However, it is important to realize that the bulk of industry's R&D investment is in the "D" and not the "R"! A number of economists, science policy experts, and government leaders have noted a shift in industrial research away from long-term basic science and toward more applied, near-term areas. Many large industrial laboratories, such as Bell Labs, have been dismantled or restructured, and in nearly all of them, the era of basic "curiosity-driven" research appears to be over. Many in the science community have bemoaned this relentless pursuit of the short-term and lament that breakthrough technologies of the future may fail to emerge in such an environment.

However, along with the dismantling of their in-house basic research, many companies and industrial sectors are forging stronger ties with universities—the repositories of basic science and the source of new scientists. Companies are finding it more profitable and reliable to scour the world for breakthrough technologies in universities, smaller companies, and national laboratories, and then license those technologies, rather than rely on their staff of in-house researchers to produce all the breakthroughs they need. Thus, the trend away from big, centralized industrial labs is less of a retreat from long-term research and more a move to outsource the research function. Where once industrial R&D was vertically integrated—with every step from idea to product taking place under one roof—now industrial R&D is becoming distributed among numerous players. Basic science is becoming a commodity.

This trend has important implications for the careers of young scientists. In the past, a young scientist could look to a large company or a national laboratory for the best facilities and most secure employment. Today, many smaller companies and start-ups are leading the technological revolution. They are nimble, focused, and fast-paced. The rewards of working in a smaller company can be staggering, especially if the small company gets much bigger or is bought out by a large firm. For example, two out of three employees at Qualcomm, a telecommunications company, became millionaires in the course of a single year. However, the success rate for most technology ventures is not high. Many more stall before they reach a big pay-out. To thrive in such a dynamic environment, scientists must remain flexible, versatile, and well-connected.

I remain seriously interested in a career in academia. Are such careers possible today?

Absolutely! Academia remains one of the principal career goals of young scientists, even though most Ph.D. scientists do not end up there! In 1995, only 46% of the Ph.D. scientists and engineers in the United States worked in academia. Today that number has fallen further. Furthermore, of those who do work in academia, only a small fraction have jobs in research universities. Many more work in a very diverse set of environments, from small liberal arts colleges to junior and community colleges. There will always be opportunities in academia, but the number may be highly field-specific.

Academic employment faced a number of pressures in the 1990s, and will continue to do so in the future. Mostly, this pressure is due to money. Colleges and universities continue to be under financial pressure to cut costs and slow tuition increases. The recent economic revival in some states has permitted funding increases to some state colleges and universities, but after years of budget freezes many schools find themselves using the new money to fill gaps created during the lean years. As a result, the number of full-time faculty positions for scientists and engineers has fallen slightly, from 173,000 in 1991 to 171,000 in 1995.

Tenure itself is under new pressures. Some schools have flirted with the abolition of tenure altogether, but most are reacting incrementally by hiring more adjunct and temporary faculty and fewer tenure-track faculty. If this trend continues, and there is every sign that it will, colleges and universities will be staffed by a few tenure-track professors and a sea of temporary or non-tenured staff.

There are also new mandates on institutes of higher education. State legislatures and boards of regents are requiring colleges and universities to increase their focus on teaching. As a result, some schools are moving away from a research-focused agenda and more toward one that balances the roles of knowledge production and dissemination. The days in which faculty could lovingly dote over their own research with little care for students or teaching are just about over!

Because of the relentless cost pressures on colleges and universities, the trend toward hiring a greater proportion of temporary and adjunct faculty and lecturers will likely continue. In 1977, full-time faculty accounted for 88% of all science and engineering positions in academia. Today, the percentage has fallen to 79%. This will come as a disappointment to young scientists trapped in a cycle of temporary or part-time academic employment. However, for those who plan a career in industry or government, the number of opportunities for teaching a course or two may actually increase.

Finally, the recent end of mandatory retirement for college and university professors may result in slower attrition of senior faculty. While some studies indicate that the number of professors who "overstay their welcome" is small, lack of mandatory retirement may adversely affect the job supply in other ways. First, senior faculty members are more expensive and a university can afford fewer of them. Second, older senior faculty members may be unable to move into new fields of interest to students or funding agencies.

There are reports—again—that there is a glimmer of light on the horizon. The children of baby boomers, now in elementary and secondary school, are starting to hit the college scene. California, for example, expects a 43%

increase in the number of high school graduates over the next 10 years, and many of those students will be moving on to 4-year colleges. Much of the growth is expected in the west, in states that have seen a tremendous increase in population over the last decade. How these states, notorious for their aversion to higher education spending, expect to pay for this influx of new students is uncertain. In any case, it is likely that more college professors will be needed to teach them.

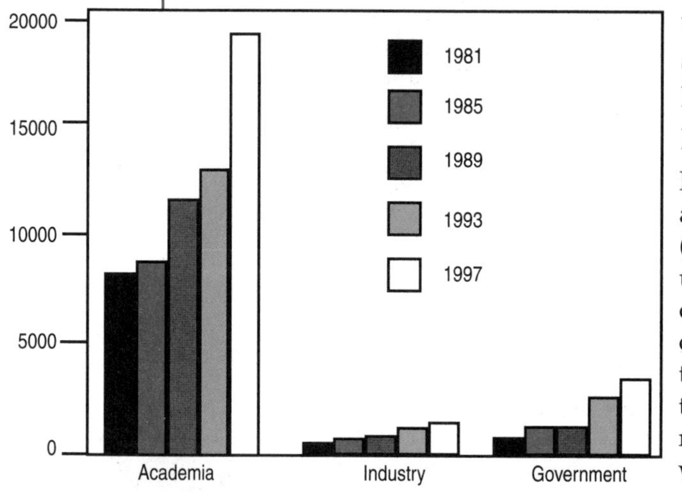

Figure 4. Growth in the proportion of new Ph.D.s taking postdoctoral positions.

What are the trends in postdoctoral positions? How long can I expect to remain a postdoc?

Postdoctoral appointments have always been around in science (heck, my Dad did one). But until recently the issue of postdocs has been shrouded in obscurity. Few meaningful statistics were gathered on postdoctoral populations, employment rules and compensation levels were poorly and inconsistently regulated, and no national organizations existed to speak to the needs of postdocs.

The number and percentage of new doctorates going into temporary postdoctoral positions immediately after graduate school has grown substantially over the last 20 years. In the past, postdoctoral appointments lasted only 1 or 2 years. Today, they can stretch on for as many years as it takes to get a Ph.D. In some disciplines, such as the biological sciences, many Ph.D. graduates and their advisors consider it "normal" to be employed as a postdoc for 4-6 years. In many cases, postdocs are in a holding pattern, building up publications and research portfolios that will make them competitive for permanent positions. As we will discuss in Chapter 7, postdoctoral work can help a young scientist sharpen and broaden his or her skills, professional network, and competitiveness. However, there is a limit. Statistics show that the probability of transition from postdoc to faculty member drops substantially after 4-5 years as a postdoc.

The large supply of new Ph.D.s, continued cost pressure on research organizations, and the growing "acculturation" to the postdoc process suggest that we will continue to have a large number of postdocs in the future. There are some reform initiatives now underway to improve the plight of postdocs—improving benefits, employment status, workplace rights, and professional development opportunities. However, no matter how "nice" the postdoctoral experience may become, many young scientists remain frustrated that they must work for so long before they can compete for a permanent job in research. Indeed, in the "good old days," young scientists used to build their publication and research portfolios while employed as assistant faculty members. Today, new assistant faculty members have as much experience as a "tenureable" faculty member had 20 years ago!

Beyond science, what general job trends should I watch out for?

Many of the changes in industry, academia, and government science employment reflect broader changes in the nature of jobs and job development in the United States and the rest of the world:

- The push toward globalization is rewarding workers who come from multilingual and multicultural backgrounds.
- Job mobility is increasing. People are not only switching jobs more often but moving from field to field more frequently.
- Free agency is on the rise. More individuals are working on their own or as independent consultants. Those in organizations are increasingly being evaluated on their individual impact on the organization.
- Technology is everywhere. Workers with technology skills continue to be highly valued here and abroad.

What about Ph.D. supply?

Enrollment in science and engineering Ph.D. programs peaked in 1993 and has been dropping ever since. News of a difficult job market is finally propagating back into the population of undergraduate science majors and many are choosing greener pastures than graduate school. Enrollments for 1997, the most recent data that is available, show an 11.5% drop overall. Astronomy is down 22%, physics is down 25%, geosciences are down 10%, math is down 26%, and chemistry is down 4%. Biology, on the other hand, is up 2%, with big increases in cell biology, genetics, and pathology.

Much of the growth in the total number of Ph.D.s produced over the last 15 years has been among noncitizens. Despite an increase in the number of overseas Ph.D.-granting universities, the United States remains the "OPEC" of graduate education. In general, foreign-born graduate students are happier with the graduate school experience than U.S. citizens, and some recent studies (e.g., Levin and Stephan, 1999) show that foreign-born scientists in the United States produce a disproportionately high number of ground-breaking scientific discoveries. High-tech industries and universities are now calling for further increases in the number of visas for foreign nationals to study in the United States, and there is every indication that the trend toward an international population of graduate students will continue.

Workplace Basics

The Essential Skills Employers Want...

summarizes a recent exhaustive survey by the American Society for Training and Development along with the U.S. Department of Labor on the skills most desired by employers. They list the following:

1. Learning to learn—the ability to absorb, process, and apply new information quickly and effectively
2. Reading, writing, and computation
3. Communication—the ability to communicate and listen effectively
4. Adaptability—Creative thinking and problem solving
5. Personal management—Self-esteem, motivation/goal setting, and career development/employability
6. Group effectiveness—Interpersonal skills, negotiation, and teamwork
7. Organizational effectiveness and leadership

Despite calls for limits to Ph.D. production on the part of some policy makers, academia is unlikely to ever adopt any significant controls on Ph.D. production. Ph.D.s are the most cost-effective means of producing scientific research. While the supply of eager U.S. citizens may be waning somewhat, the demand for higher education from noncitizens is huge and is only likely to grow.

Summing It All Up: The Scientist of the 21st Century

The 6+ year lag time between enrollment and graduation in a Ph.D. program ensures that new Ph.D. graduates will encounter a job market that is significantly different from the one they inhabited when they decided to go to graduate school in the first place. One theme this book will return to again and again is the fact that young scientists have a huge range of career opportunities in front of them if they are willing to consider their training more broadly. Ph.D. scientists and engineers are EVERYWHERE in today's economy; in law, business, government, the non-profit sector, and the entertainment industry. While the health of the job market in each of these industries will wax and wane with the overall health of the economy, there will always be a premium placed on bright, creative, hard-working individuals. The challenge for many young scientists is to understand how their skills and training translate into opportunities outside of the ivory tower...

Summary

- As a scientist you need to be aware of the larger trends shaping funding and employment.
- Federal funding for R&D will grow, but not as fast as industrial funding.
- Postdoctoral appointments will continue to grow and more young scientists can expect to spend at least a few years as a postdoc.
- Ph.D. production will continue to climb, limited only by federal funding and the supply of interested students.

More Data!

For those of you with an unquenchable thirst for employment data, the past five years have seen a tremendous expansion in the range and depth of employment data for scientists.

- The Commission for Professionals in Science and Technology (www.cpst.org) publishes employment and salary data on many fields of science, including a recent study of trends in postdoctoral employment.
- The American Association for the Advancement of Science (AAAS) has aggregated years of federal and state R&D funding data in comprehensive, easy-to-read graphs and tables. Check out their guide at www.aaas.org/spp/dspp/rd/guide.htm.
- The biggest supplier of statistical information on science education and employment remains the National Science Foundation's Division of Science Resources Studies. Their biennial Science and Engineering Indicators is a compilation of the current data on everything from science education to graduate enrollments and employment. The 1998 S&E Indicators Report can be found at http://www.nsf.gov/sbe/srs/seind98/start.htm.
- Eric Weinstein and colleagues at Harvard have undertaken a major study of the economics of graduate training in the sciences. Their site, the Project in Economics of Advanced Training (PEAT), has some thought-provoking analyses of the forces and policies shaping graduate education. Check out: http://users/nber/org/~peat.

Many individual professional and scientific societies are now compiling their own statistics on graduate education and employment. Check out studies from the:

- American Psychological Association: http://research.apa.org/96-97insert.html
- American Geophysical Union:
  http://www.agu.org/eos_elec/eosfisk.html
  http://www.agu.org/sci_soc/cpst/employment_survey.html
- American Institute of Physics: http://www.aip.org/statistics/
- American Chemical Society: http://www.acs.org/careers.html
  and many more... check out your own professional society.

AJL
You have to be tough to survive on this desolate planet. What did YOU do to end up here?
I got a PhD in Mathematics. You'd be surprised how often it comes in handy in a place like this.

Now the Good News
The World of Opportunity Open to Scientists

2

Young people themselves don't realize how valuable they are with a Ph.D. It means an ability to think deeply, solve problems, analyze data, criticize, and be criticized. [Ph.D.'s] often don't realize the breadth of what they are capable of doing.

Neal Lane
Science Advisor to the President of the United States

Having read the last chapter you might be feeling rather pessimistic about a career in academia. Opportunities are limited, competition is fierce, and the rewards of such a career path might take a long time in coming to fruition. You may be inclined to throw up your arms in despair, quit your scientific career, and begin a new life as a short-order cook or cab driver! Do not despair. The career opportunities in research science may be hard to come by. But, the opportunities that exist for people trained in science and engineering are HUGE outside the ivory tower. In some ways, there has never been a better time to be a techie.

As outlined in Chapter 1, careers in the United States are changing, with an increasing emphasis on independence, versatility, and mobility. Employers in the new millenium are seeking individuals who are independent problem-solvers, quantitative thinkers, and articulate communicators. They are seeking individuals who have carried out complex projects, overcome obstacles through creative thinking, and operated with a minimum of supervision.

In other words, they are seeking people like you.

The same skills that make you a good scientist would also make you a good business owner, non-profit fund raiser, administrator, banker, lawyer or public policy analyst. The skills that are of MOST value to the outside world are not the ones you might associate with a scientific career. While your particular aptitude for mathematics and your skills with particular devices and techniques may be critical to your scientific career, the skills that are the most valuable in the outside world are the broader, more general qualities you rely on every day to do your work. This is true even for research careers in academia: employers require MORE than just brains. They are also looking for people who have a host of "soft" skills. The combination of technical training and these soft skills is an amazingly powerful combination.

Revenge of the Nerds

Believe it or not, graduate school is actually a good training ground for developing many of these "soft" skills. We have been so channeled into thinking that graduate school is simply about science that we often forget the variety of other skills and personal qualities we have had to develop while in graduate school just to get our work done. These "transferable skills" are extraordinarily valuable; in fact, they are the hallmark of the most successful and productive people working in today's economy.

Below is a list of some of the common transferable skills that we learn in graduate school. As you read though this list consider all the ways in which your years as a scientist have given you experience in developing these skills:

Transferable Skills

- ability to function in a variety of environments and roles
- teaching skills: conceptualizing, explaining
- counseling, interview skills
- public speaking experience
- computer and information management skills
- ability to support a position or viewpoint with argumentation and logic
- conceiving and designing complex studies and projects
- implementing and managing all phases of complex research projects and following them through to completion
- knowledge of the scientific method to organize and test ideas
- ability to organize and analyze data, to understand statistics, and to generalize from data
- combining and integrating information from disparate sources
- critical evaluation
- ability to investigate using many different research methodologies
- problem-solving
- familiarity with the committee process
- ability to do advocacy work
- ability to acknowledge many differing views of reality
- willing to suspend judgment, to work with ambiguity
- ability to make the best use of "informed hunches"

We tend to take these skills for granted. In fact, most people who are unfamiliar with a career in science have no idea that graduate school develops skills like these. For example, many people are unaware that scientists do as much public speaking as we do. Our experience giving talks at national and international meetings, as well as our teaching experience, exposes us to public speaking in a variety of settings. People unfamiliar with a career in science probably don't realize that.

Many people unfamiliar with science also fail to appreciate how much creativity, ambiguity, and idealism is part of our work. In fact, most non-scientists tend to think of us as cold, aloof analytical types who are more comfortable with numbers than people (this is a subject we'll return to in Chapter 13). You know that science is full of passion, commitment and emotion. Most of the outside world think of us as Mr. Spocks. We know that we're really more like Captain Janeway or Picard!

This is not to say that our technical aptitude is not valuable—it is. Computer skills and Internet literacy are becoming bedrock requirements in many workplaces. We scientists have a huge head start. After all, the Internet was conceived and designed by and for us. Starting at the earliest levels of education, we have used computers as tools for computation, communication, and analysis and many graduate students end up developing powerful software, working with complex data collection and storage issues, and designing new devices. We have a level of familiarity and comfort with information technology that the rest of the world is frantically trying to acquire.

You're Smarter than You Think!

In addition to transferable skills, a training in science tends to develop some very valuable personal qualities. For example, it is likely that you are very, very smart. You probably did extremely well in college; maybe you were the top in your department. But when you joined all those other smart people in graduate school, you probably ended up feeling rather average. In such a peer group, it is natural to feel only average! Don't forget that, to the rest of the world, you are a rocket scientist.

Surviving a rigorous science curriculum required not only brains but stamina as well. Let's face it: you probably work really hard. You're used to working hard, you may even LIKE to work hard. Again, you may consider this to be normal but this is only because you live in a very extraordinary peer group. In graduate school, brains and hard work are requisites. In the outside world, these are the qualities of a high achiever.

Consider the list of personal qualities below. Think about all the times you have had to rely on these skills to move ahead in your science career...

Personal Qualities

- intelligence, ability to learn quickly
- ability to make good decisions quickly
- analytical, inquiring, logical-mindedness
- ability to work well under pressure and willingness to work hard
- competitiveness, enjoyment of challenge
- ability to apply yourself to a variety of tasks simultaneously
- thorough, organized, and efficient
- good time management skills
- resourceful, determined, and persistent
- imaginative, creative
- cooperative and helpful
- objective and flexible
- good listening skills
- sensitive to different perspectives
- ability to make other people "feel interesting" (like your advisor)

Some of the items on this list seem rather obvious, while others might surprise you. For example, you might not think that you have good time management skills. But to complete a thesis while carrying out teaching responsibilities, looking for a job, and submitting papers for publication, effective time management is critical.

They Didn't Tell Me I'd Learn That!

Twenty-two scientists at various stages of their careers were asked the following question. "Of the many skills that people develop while in graduate school, which are the most valuable in the outside world?"

The top five answers were:

1. Ability to work productively with difficult people
2. Ability to work in a high-stress environment
3. Persistence
4. Circumventing the rules
5. Ability and courage to start something even if you don't know how yet

Other items on this list might surprise people in the "outside world." For example, people unfamiliar with research science have no clue about the level of cooperation and community service that is a part of a scientific career. They don't know that we review papers; edit journals; and provide data sets, programs, equipment, and analyses all for free. We do this because it is part of the cooperative endeavor of doing science. This communitarian ethic is one of the most powerful and positive aspects of our community.

The outside world may also fail to appreciate the role of creativity in our profession. In fact, creativity is the hallmark of the best science. The ability to "think outside the box" has led to some of the most important breakthroughs in science, and scientists continually strive to approach difficult problems from new directions. The outside world may not think of creativity as a skill needed in science, but we know that it is essential.

The Bottom Line

The lists of transferable skills and personal qualities above show that many of the abilities we rely on every day to do our jobs are very valuable skills when applied in the outside world. There is one added benefit: people who excel at these things also tend to be very well compensated in the outside world! In fact, many of these skills are the hallmark of leadership potential. Some firms have realized the hidden value of a Ph.D. or M.S. and have begun to explicitly recruit scientists and engineers from top universities. Starting salaries over $100,000 are not atypical for fields such as management consulting or investment banking.

As scientists we have grown used to the idea that our work will be intrinsically, but not necessarily extrinsically, rewarding. In fact, we are suspicious of large extrinsic rewards, such as high salaries. Many scientists, especially those who have known no other profession, tend to ridicule large paychecks as "selling one's soul." Many believe that in order to achieve such handsome compensation one has to surrender one's independence and intellectual vitality. This is simply not so. It is true that high caliber talent tends to be rewarded far more handsomely in the outside world. But many of those jobs have with them a set of intrinsic rewards, intellectual challenges, and independence comparable to that in academia.

The Curse of the Ph.D.

Having gotten to this point you might be thinking that a science training is the best education in the world. With a list of skills and personal qualities such as those above you might be wondering, Hey, why aren't we running the whole planet by now?

> *The key is not how much intelligence you have, but the type of intelligence, and your willingness to use it. Steven Jobs was a back alley hustler with street smarts, not an intellectual. His "genius" was the ability to dream dreams, then connive and cajole others into cooperation.*
>
> Joseph V. Anderson
>
> *Weirder than Fiction: The Realities and Myths of Creativity*

While scientists have extraordinary training, skills, and abilities we are far from perfect. We are also taught a number of habits and beliefs that are not only detrimental to success in the outside world, but even hinder our success—not to mention happiness—in science. Succeeding in any career requires not only brains but business acumen, good people skills, and vision. And it is in these latter categories that the science education falls short.

Because so many academics eschew the private sector (and thrive in a semi-socialist system of government grants), few of their students get meaningful exposure to life in the "real world." As a result it is easy for young scientists to leave school without much understanding of the value of risk, the value of time, the importance of people skills, or a concept of vision; these four skills are extremely important to career success in and out of science.

Risk Aversion

Risk aversion is more than just the tendency to avoid risk; it is the inability to weigh risk and reward and a failure to recognize when prudent risk taking is needed. The career of science can be very attractive to risk averse people! From a college grad's perspective, a science career can seem a very secure pathway: hard work seems to be rewarded with tenure and security. Once in grad school, many students find that the financially stressed, competitive world of research science actually promotes intellectual conservatism and risk aversion. Research groups make safe, incremental research steps because that is the only way to get funding. Few researchers can get grants nowadays for proposing a wacky idea outside their sub-discipline. And students learn this lesson fully. Have you ever seen someone try a daring new project for his or her thesis, fail to make it work, and still get a Ph.D.?

The Value of Time

A failure to understand the value of time is a second "business skill" that is critical in the real world but remains an utterly alien concept in graduate school. Most of this stems from the fact that academia is an environment steeped in penury. Not only are most graduate students paid little more than the minimum wage, but PIs are often forced to pinch pennies to ridiculous degrees. In this environment it can seem sensible to spend a week of a grad student's time to build or repair a device that would cost only a few hundred dollars to replace. Once one is out in the real world, one learns that in many cases the time spent pinching pennies simply does not pay off. In the working world people's time is much more expensive, and decisions and actions lose their value the longer they are delayed. Graduate school teaches one to be careful and meticulous. In the real world, decisions often must be made with insufficient data because of time constraints.

Advice from the Field

We asked representitives of business and industry what they thought of the preparation of science and engineering Ph.D.s. The employers told us that ... they find shortcomings in three areas: communication skills, including teaching and mentoring; appreciation for applied problems, particularly in an industrial setting; and teamwork, especially in multidisciplinary settings... Here is a typical comment ... "If you walk on water technically but can't explain or promote your ideas and your science, you won't get hired."

Phillip A. Griffiths

Chair, Committee on Science, Engineering and Public Policy, National Academy of Sciences

Poor People Skills

It is truly ironic that so many professors, hired to communicate and deal with people on a daily basis, have poor people skills. They can be insensitive, overly critical, or shy and socially inept. Again, this may reflect a self-selection effect: shy, anxious, or obnoxious people may find it hard to get ahead in the real world but not so in academia! Not only do graduate students receive no training in dealing with people (many of you will become managers someday, really!), but also because of the heavy intellectual bent to graduate school, other forms of intelligence are greatly undervalued. Let's face it: grad school is a "book-smart" culture. Sense of humor, tact, joviality, and empathy are all aspects of emotional intelligence that are rarely discussed, never taught, and patently undervalued!

Thinking Big Picture

As a scientist you know that insight is often gained by having a breadth of vision and seeing connections that have not been recognized by others. The same is true in the real world. However, it is startling how infrequently graduate programs try to instill in students a sense of vision either in their research or their science. With the risk aversion also comes a narrowness of view that encourages sub-subspecialization. Graduate students are told to focus, focus, focus. "Eliminate those distractions!" "Don't take those outside courses!" "Just get your work done!" As a result, students learn a nose-to-the-grindstone style of thinking that is more pervasive in graduate school than ever before.

While success in graduate school does require some focus and diligence, an absence of ANY external stimulation, macroscopic focus, or grand vision can lead to stagnant, derivative research (not to mention a lackluster career), as well as tremendous myopia about the world around you.

What can you do to improve the weaknesses of your science training?

Risk aversion	Learn to understand risk and reward. Practice identifying and quantifying it.
Valuing time	Determine what an hour of your time is worth, then calculate the pay-off for various activities.
Poor people skills	Join ToastMasters (www.toastmasters.org) or some other social group to practice social skills.
Thinking big picture	Develop a discussion group in your department. Invite outside speakers. Discuss way-out ideas such as colonizing Mars, the politics of cloning, cyber-biotechnology, or entrepreneurship.

Presenting the Best Side to Employers

The challenge you face in exploring careers outside of science is partly one of educating prospective employers about the true range of your skills and abilities. Most people unfamiliar with science simply do not know the range and breadth of what we do. Many cannot easily make the logical jump between a successful career in science and a successful career in a non-science area. Some harbor negative stereotypes about science and scientists. Your challenge as a job seeker is to educate and enlighten prospective employers about your true talents, and to reinforce positive stereotypes and dash negative stereotypes about scientists.

Some Who Have Made the Change

Many of the people who have received science Ph.D. and Masters degrees have gone on to careers that do not utilize their specific scientific training. Hard numbers are difficult to come by because there has been no systematic attempt to track the employment of scientists who pursue alternative career paths. However, the *Survey of Doctoral Recipients* shows that the number of Ph.D.s employed in "Business/Industry" and "Other Employment" is growing. In 1977 these two categories held 34.8% of all those who were 5–8 years out of their Ph.D. program. In 1991 this number had grown to 45%. However, it is impossible to know how many of the people in these categories were in non-science careers.

In the past, people who left research science vanished entirely from the scientific community. They were no longer seen at meetings and they no longer published papers. Often their advisors felt disappointed or even betrayed that they failed to follow in their footsteps and so they became the prodigal children of science.

Today, these black sheep are some of the stars of the new economy. Several books and scores of articles have appeared in the last 2 years featuring the stories of scientists who have made the transition to a wide variety of career fields. Scientists are EVERYWHERE!

In 1996 we featured a handful of young scientists who had made successful transitions to careers beyond the bench. Today there are many more. In the next section we check in with some of those alumni from the first edition of this book and meet some new faces. While their motivations and career paths are different, they all share a common experience as former graduate students. Each one has found that the skills and traits they developed in graduate school have broad application in the real world.

When we checked in with **Dr. R. Brooks Hanson** (Ph.D. Geology, UCLA, 1986), in 1996 he was a senior editor at *Science* magazine. Today he is still at *Science*, but he has a new title: Deputy Managing Editor. While in graduate school, Hanson undertook the "ideal" geological project, incorporating laboratory experiments with computer modeling and extensive field work in the picturesque White Mountains of California. After a 1-year postdoc at the Smithsonian, Hanson took a job with *Science*. Today, he oversees physical science for the magazine and several Ph.D.s-turned-editors. "We look for people with broad interests and backgrounds, and an exposure to cutting edge science," explains Hanson. According to Hanson, many of the best candidates come from top 10–20 departments in their field. "We feel we can teach editing, but not so much scientific knowledge or judgement," he explains. Now in a management position, Hanson laments not having more formal training in business and management. "Science (not only the magazine) is increasingly a business, and anyone who is successful will eventually be involved in managing others. I think the absence of such training makes it much harder for a young scientist to set up and run his or her own lab."

Dr. Craig Schiffries (Ph.D. Geology, Harvard, 1988) has pursued a career that spans several sectors. He majored in geology and geophysics at Yale University and in philosophy, politics, and economics at Oxford University. Schiffries received his Ph.D. from Harvard in 1988. He has spent the last 10 years in Washington, D.C., in a variety of science policy positions. As a Congressional Science Fellow in 1990-1991, he served on the staff of the Senate Judiciary Committee, where he worked on legislation to ensure that federal laws keep pace with changes in technology. According to Schiffries, scientists are poorly represented among those who are formulating the local, national, and international governmental responses to these changes. Schiffries then joined the American Geological Institute (AGI) as their first Director of Government Affairs. He subsequently served as Director of the Board on Earth Sciences and Resources of the National Research Council/National Academy of Sciences (NRC/NAS), where he headed a team that completed numerous studies on issues related to geoscience and public policy. Schiffries believes that a stint in public service in Washington, D.C., is an extraordinarily enriching experience for a scientist and is of great value to the scientific community and the nation. This year, Schiffries returned to his alma mater, Yale University, where he is a visiting professor. He's even teaching one of the classes he took as an undergraduate, in the same lecture hall! This summer Schiffries will move to the private sector by joining the Monitor Group, an international strategy consulting firm.

Dr. Randy Krauss (Ph.D. Microbiology, University of Alabama at Birmingham, 1992) knew he wanted to move away from the lab, but to what? Part-way through his first postdoc, Randy realized that he'd have to take a second postdoc to obtain an academic or industrial job. At the same time he realized that he was more interested in explaining science to the lay public than grinding away at the bench. Seeking to satisfy his interests in both communicating science and new educational technologies, Randy began working on a CD-ROM on biotechnology while continuing his postdoc during the day. Then one day he read about CityLab, an innovative biotechnology science program for middle and high school students in the Boston area. He sought out the organizers of the program and convinced them to make him Laboratory Coordinator. Today, Randy continues to work at CityLab and is an instructor at Boston University's School of Medicine. His advice to other Ph.D.s interested in careers in education: "Try it out. Working with kids is not easy!"

Today, **Dr. Loren Shure** (Ph.D. Geophysics, Scripps Institute of Oceanography, University of California at San Diego, 1982) is Director of Signal Processing and Applied Math at Mathworks, the makers of MATLAB and several other successful software products. Shure had finished a postdoc at Woods Hole Oceanographic Institute when she finally decided that she wanted a broader experience than that of research science. Having extensive experience and interest in developing software for geophysical data reduction, Shure made inquiries to a number of small software firms, ending up at Mathworks. She was the first employee. In 1996, when we first met her, Mathworks had 87 employees and Shure managed a team of six. Today, Mathworks employs over 500 and Shure runs a group of more than 25. The fast pace of product development and the teamwork environment at Mathworks were two refreshing changes from the academic research setting, according to Dr. Shure. In addition, Shure finds herself playing a large mentoring role in the organization. "It's like being a thesis advisor," she explains, "and is one of the most satisfying aspects of my career."

Too narrow! That's what **Dr. Signe Holmbeck** (Ph.D. Chemistry, University of Wisconsin at Madison, 1994) realized about her life as a postdoc at The Scripps Research Institute (TSRI). Having studied protein-DNA interactions using NMR spectroscopy for over 3 years, Holmbeck decided to explore as many other avenues as possible while still carrying out her research. She joined and later led the postdoc organization at TSRI and expanded the lecture series to include more speakers from industry. Having built a strong network in the San Diego biotech industry she and two other other graduate students began a technical consulting group to help biotech investors evaluate promising technologies. Through that experience she learned about patents and patent law. Wanting to solve her own two-body problem, Holmbeck explored patent law careers in the San Francisco Bay area and landed a position as a "patent agent trainee" with a law firm in San Jose, California. "I LOVE what I'm doing," says Holmbeck. "I see a much broader range of science and technology and I am making great use of my technical background and logic skills." Holmbeck's workplace has many other Ph.D. scientists—with degrees ranging from theoretical physics to pharmacology. "It's great to 'not be alone'" Holmbeck's advice for other postdocs who want to explore other options besides research: "Have the courage to be among the least productive researchers in your lab for a while. The time you spend researching your career options will be time well invested."

When we last saw **Matthew Richter** (Ph.D. Applied Physics, Stanford University, 1993) he was a staff scientist at Burleigh Instruments, Inc., in Rochester, New York, working on the company's ultra-high vacuum scanning tunneling microscope (STM). Then 2 years out of his Ph.D., Richter had switched from pursuing an academic career to a career in industrial design and manufacturing. He found Burleigh Instruments by combing

through *Physics Today Annual Buyer's Guide* and calling every STM manufacturer to see if there were positions available. To avoid being shuffled off to the personnel office, he would ask to speak to an engineer about a technical question. Once the connection was made, he would admit his scam and inquire about potential job leads! Richter found that his aggressiveness in making call after call gave him a distinct advantage during his search, and he urged other physicists not to be intimidated about doing the same. Today, Richter is a veteran of the fast-paced, hi-tech workforce. He has worked for four different companies since Burleigh Instruments designing and marketing a variety of instrumentation for the semiconductor industry and is currently working out of his home in San Jose, California, as the West Coast representitive of On-Line Technologies. On-Line Technologies is a manufacturer of infrared spectroscopic instruments, based in Connecticut. When asked about the many job changes he has made, Richter explains that "in the world of hi-tech startups the best way to move up is often to switch companies. You gain a new perspective and new technical skills with each move. In the Silicon Valley, such rapid career change is typical." In talking to other Ph.D.s, Richter has found that the fear of switching jobs is just like any fear of change. "It takes some practice, but if you are willing to work hard there is always an opportunity."

Stacey Schultz-Cherry (Ph.D. Cellular and Molecular Pathology, University of Alabama at Birmingham, 1995) loves her job. She is a research scientist at the U.S. Department of Agriculture's (USDA) Southeast Poultry Research Laboratory in Athens, Georgia, a job that she feels gives her all the joys of doing science with fewer of the headaches. Her only complaint: "I don't like chickens!" In grad school, Schultz-Cherry was on the academic track, following in her father's footsteps. The closer she got to finishing, the more she realized that she loved the bench and didn't want to become a manager. She discovered the world of emerging infectious diseases and the opportunities that exist in government laboratories. "I get to play in the lab all day... and I don't have to write grants!" To keep her hand in academia she maintains an adjunct faculty position at the University of Georgia, where she is involved in teaching and mentoring graduate students and postdoctoral fellows. Schultz-Cherry's work does not only involve lab work: she talks with USDA officials, members of the poultry industry, and Congressional staff. "Everything I learned to do in grad school and in my postdoc: managing, budgeting, recruiting have become critical skills for my present career. It really is a wonderful job...in spite of the chickens!"

You Could Be One, Too

There are thousands of other science Ph.D. and Masters recipients who have left the world of research science for the larger world around them. Not every story is a success and not all career transitions have been easy.

But overall, these prodigal scientists have discovered that their scientific training and experience have served them well in the outside world. If they can do it, why can't you?

Summary

- Your training has given you a range of transferable skills and personal qualities that are extremely valuable in a wide variety of work settings.
- The national shift in employment towards highly skilled, motivated, independent workers favors those with a training in science.
- Many individuals who trained in the sciences are pursuing fulfilling careers outside of research and academia.

Spare change, man?
I fear change.
CLINK
AJL

The Science of Change

3

Change is avalanching upon our heads and most people are grotesquely unprepared to cope with it.

Alvin Toffler
Future Shock

By now it is probably painfully clear to you and your colleagues that science is changing. Not only are the employment opportunities for scientists changing but the mechanisms for funding, conducting, and publishing science are changing as well. While these changes are affecting people at all levels in their careers, the uncertainty, frustration, and fear they produce are particularly acute for those who are just establishing themselves.

You would think that scientists would be particularly good at dealing with change. After all, we work in a profession that thrives on discovery, innovation, and new ways of doing things, right? Scientific disciplines are periodically wracked with radical change, such as the discovery of the electron, the theory of evolution, plate tectonics, and the big bang. We love change, right?

Actually, if you look at the track record of science you will find that we do a poor job of coping with change! Scientific theories take years to become accepted—not because it takes that long for proof to be produced—but because it takes that much time for scientists to change their minds and accept them. Some never do!

In the same way, scientists can become stressed out by the prospect of change in their careers. As you know, a scientific career involves many years of training, commitment, and apprenticeship. Decisions you make in your first or second year in graduate school lock you into a specialization that lasts not only through your graduate education but possibly through the rest of your professional life. Being forced to confront change can feel devastating, as if you're on a train that has suddenly lost its tracks.

But why can career change be so difficult for scientists? I think there are several reasons. First, you have made a huge investment of time and energy to get where you are. Facing an unplanned change—such as changing jobs, changing fields, or leaving research altogether—can seem like a huge loss of investment. Many people feel a sorrow not too different from that experienced when losing a loved one. Second, change can often mean losing one's professional identity. It was easy to describe yourself as a scientist. You knew what the label meant, and so did everyone else. But in undergoing a career change, what do you call yourself? And what do other people think about you? Finally, there can be a sense of embarrassment or even shame at not having managed your life better.

Let's get one thing straight: nobody likes change.

Change is painful, uncertain, and requires extra work. Change involves confusion, fear, anger, frustration, and sadness—not exactly the kind of emotions we enjoy experiencing day after day. However, change also brings new opportunities. New ecological niches open up. Nature has a double-edged way of reacting to change: some species become extinct and some new species evolve. You are confronted with a similar choice: are you a mammal or a dinosaur?

People experience change throughout their lives. However, men and women in their 30's commonly encounter a difficult period of transition: one that often involves their professional life. The 20's are often described as a "novice period" in which you are learning and developing but have not yet gained full independence. Once you have gained independence from graduate school, a number of deep questions rise to the surface, such as "What do I really want to do with my life?"

Becoming a Change Master

The process of personal or professional change is made up of a series of fairly predictable steps. While people may spend more or less time at each point, they ALL go through the steps sooner or later. Most people go through the process without any awareness of what's happening to them. As a result, change can be frightening and disorienting. Understanding the entire transition process before you go through it can help as an emotional catalyst: it can lower the "emotional activation barrier" of the process and can speed the transition to your new self.

The Transition Curve

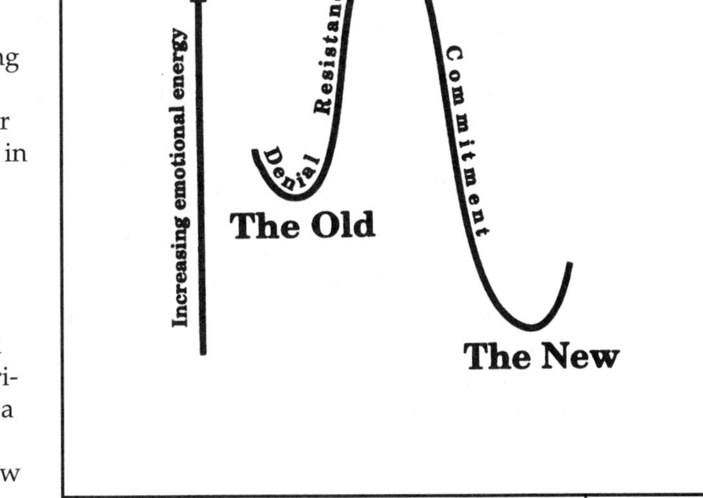

In *Managing Personal Change: A Primer for Today's World,* authors Cynthia Scott and Dennis Jaffe present their "transition curve" as looking something like a potential energy well. They label four parts of the curve as stages in the process of change: Denial, Resistance, Exploration, and Commitment.

I have found it more useful to think of change as a barrier to overcome rather than a pit to wade through. The curve to the right is my view of the change process, with various parts labeled in the same way as Scott and Jaffe's curve. Viewed in this way, the process of change is like a chemical reaction. To move through the change process we must traverse a barrier that is made up of our own anxieties and fears, as well as the fear and anxiety of those around us. The diagram implies that this change is irreversible. In the case of changing from a research science to a non-science career this may not be strictly true, but the barriers to returning to a career in science grow enormously with time.

Denial

Denial surfaces in two ways during the process of change. First, you can be in a state of denial, unable or unwilling to see the evidence of change around you. People often construct their lives and behaviors subtly to avoid reminders of the change that may be looming. One postdoc I know completely burrowed into his work in his final year. His strategy for avoiding the painful change to a new job was to ignore it. Jobless, he eventually had to leave the United States.

The other way denial rears its ugly head is by way of the reactions of others. Quite a few faculty members in my department used to dismiss the dismal job situation by blithely saying, "if you're the best, you'll get a job." This was not true at all! However, they were unable or unwilling to see the clear evidence around them that their students, even "the best," were having a hard time finding work.

You might also hear statements like:

The job situation is cyclical and all this will pass.
How bad can the job situation be? After all, ___ got a good job last year.
This doesn't affect me – I have a job.

It can be terribly discouraging to hear these things from those around you. In the particular case of advisors and supervisors, denial can be downright frightening, because disagreement or complaint on your part could be interpreted as a "bad attitude." However, even though your advisors, teachers, or peers may be in a state of denial, you can't afford to be. Ignoring the forces that are affecting your career may allow you to operate as if nothing is wrong… for a while. But while in denial you may ignore potential opportunities for change and only heighten the magnitude of the crisis that is to come.

So what do you do to counter denial, either in yourself or in others? For starters, it is important to challenge your assumptions and those of the people around you. Why do they believe what they believe? Is their opinion based on data or conjecture? Do they have a bias? Do you? Another useful approach is to simply gather data. Talk to those around you. Look at reports and articles. Is the job situation as good as they say? Is the job situation as bad as you fear? Most importantly: TALK TO PEOPLE. People with unbiased perspectives can give you a rational view of the situation and help you understand the forces of change that abound.

Of course, you can remain in denial as long as you like, operating in the old model and rejecting the new. However, as things change you will be increasingly at odds with the rest of the world. Psychologists call this a state of "cognitive dissonance." Eventually you will face an undeniable crisis: you can't find a job in research, or—if you are lucky enough to land an assistant professorship—you can't find any decent students to admit to your Ph.D. program. If you've waited this long to deal with the problem the process of change will likely be extremely difficult and costly. The choice is yours.

Resistance

These young scientists are just a bunch of whiners.

Resistance is probably the least pleasant part of the change process. It involves anger, sadness, frustration, and depression. It is the period in which the magnitude of the problem facing you is clear, but the solution remains out of sight. You may hear yourself, or people around you, saying:

_________ *is out to get me.* (fill in the blank: my advisor, the department chair, etc.)
This job situation is unfair.
These people have a bad attitude.

The symptoms of resistance are similar to those of mild depression:

- You are unproductive.
- You have low energy.
- You are being careless.
- You lack enthusiasm for work.

What can you do about this? For starters, it is important to take it seriously. Your feelings of sadness, loss, and depression are legitimate, so acknowledge them. Meeting with a counselor or even just talking to a friend can be a great way of dealing with your feelings of loss and frustration. While career counselors are familiar with helping people through this difficult period, scientists facing what they perceive as the total destruction of their way of life should seek counselors with experience in grief and crisis management. Depending on your current institution or employment, good counseling may be available without charge. Any other means of relieving your frustrations (short of committing a felony) is great, but nothing beats sharing your problems with others.

Exploration

There comes a moment when, for the first time, the negative, inward-directed energy you are dealing with begins to break out into a new form. It is the point at which your consciousness decides that THERE HAS TO BE A WAY OUT! This is the moment when resistance shifts toward exploration. It is an acknowledgment that the old ways aren't working anymore. And, as the transition curve diagram suggests, it's the point at which you begin to reclaim some of the energy that you expended in getting to this stage. At this point, people begin asking questions of themselves and their surroundings. While no clear avenue is in sight, the acknowledgment of alternate pathways becomes important. You may be hearing yourself ask:

Where did _______ go after she finished her Ph.D.?
What does it take to become a __________ ? (insert any non-traditional field: advocate for the homeless , city council member, artist, inventor)
What do I really like doing and where can I do it?

Presently a number of scientific organizations and professional societies are asking these sorts of questions. Some are publishing regular articles about scientists in their discipline who have launched off into different careers. These stories are popular both with students and faculty. Even your advisor might be quietly yearning for a change.

The period of exploration can also be frustrating, characterized as it commonly is by a feeling of disorganization and disorientation. You may have broken out of the old mold of thinking, but you haven't quite locked into anything new. You may find that your productivity at work has dropped or that you're spending a lot of time looking around for information or inspiration without being sure where you are going. THIS IS HEALTHY! It's impossible to know where you're going just yet, but by constantly turning over stones you can begin to find clues about where your new opportunities lie.

There are a few things you can do to make this process a bit less disorganized. One is to start a career change journal in which you capture some of your thoughts and ideas. Another technique is to block out some regular time in your work week and go to a particular place and just sit by yourself and simply think about your situtation and your options. This quiet period is very valuable in helping you sort through the things you are learning.

Commitment

The final stage of change, commitment, is both fun and scary. Finally, from all of your agonizing and exploration, a particular path begins to become clear to you. Just as in a chemical reaction, you seem to regain all that emotional energy you expended during the first three steps. In the stage of commitment you may hear yourself saying:

I have just signed up for on-campus recruiting.
Let me send you my resume.
Five years from now I want to ...

The process of commitment also represents a change in attitude. You are shifting from:

- facing a problem to gaining an opportunity
- the present to the future
- what you can't control to what you can control

A Final Note

While this change stuff is never supposed to be easy, for scientists facing an uncertain future, it can be particularly traumatic. Preparing for a research science career requires many years of hard work, deferred compensation, and training. Many consider their scientific career as much more than a job; they consider it a calling. To have all that overturned by forces that are totally beyond one's control and to be plunged into a period of financial as well as professional uncertainty would be extremely difficult for anyone. There are no easy words here. Many of your friends and colleagues are going through the same thing you are. Developing some familiarity with the process of change can not only help speed you through the process but can make you a more effective counselor, mentor, and friend. Change is inevitable; how we deal with it is what matters.

Summary

- As a scientist you need to be aware of the larger trends shaping funding and employment.
- Change, in your career and in your life, is inevitable.
- By understanding the general process by which people go through change you can make your career transitions easier.
- Those who can exploit change and adapt quickest will have the most opportunities.

Just because you're supreme ruler of the galaxy, doesn't mean you're too good for career counseling.
Yes, mother.
AJL

The Career Planning Process

How Do I Start?

4

It is all very well for author and physicist Peter Feibelman to advise young scientists to "apply the same brain power to planning their careers" as they do to their research, and not to "assume by investing your youth that you are entitled to a job." But it is something else, in the absence of help with career planning, for graduates to be unexpectedly faced with the need to reformulate job-hunting strategies on their own.

Sheila Tobias, Daryl Chubin, and Kevin Aylesworth
Rethinking Science as a Career

Scientists and non-scientists alike tend to think that career planning is the same as job hunting. Thus, most folks buy their first career planning book or take their first step in a career planning and placement center only when they are in the process of actively looking for a new job. Many march in with draft resumes already in hand: "please look at my resume for me, give me some pointers, and I'll be on my way."

In fact, you are at a significant competitive and professional disadvantage if you treat career planning as simply getting a job; it is much more.

Career planning is actually a host of professional and personal actions people take to educate themselves and the outside world about their unique talents, gifts, and capabilities. It is not an activity you start only when you are actively looking for a new job. Ideally, there are aspects of career planning that you engage in, at some level, every day. Thus, career planning is more analogous to professional development than it is to job hunting.

This is not to say that job hunting is not an important part of career planning; it is. But modern career planning puts job hunting at the apex of a number of other activities that help to prepare and strengthen you for the actual job hunt.

It may seem frustrating to be told you have more homework to do before you start applying for jobs, but it is really not that bad. Most people, when they first encounter modern career planning, go through all the steps while at the same time sending out resumes. As they go through the career planning process they usually become better job seekers.

There are serious drawbacks to ignoring the process of career planning and plunging ahead with your job search unenlightened. First and foremost, career planning WORKS! Not only does it teach you more effective ways of finding employment; it helps you become a more competitive applicant. Second, when out in the real world looking for jobs, you will be competing with other people who have gone through the career planning process. Employers will compare your job materials to other less qualified, but more polished, applicants. You run the risk of being a diamond in the rough that loses out to a well-cut cubic zirconia!

Most importantly, the process of career planning will also help you to become a better scientist.

The skills and activities that are a staple of career development—self-assessment, networking, and researching opportunities—are important survival skills for scientists. In many cases, the process of keeping your head up for new opportunities can lead you along novel pathways of research and development. These pathways can also turn up new sources of money, perhaps one of the most critical survival skills these days.

The Career Planning Process

Stanford University's Career Planning and Placement Center puts the diagram that follows in all their introductory career planning and job hunting materials. The Career Planning Pyramid is broken up into four levels: ***Self-Assessment, Exploration, Focusing,*** and the actual ***Job Search***. The diagram illustrates that the activities we traditionally think of as "looking for a job" actually are the pinnacle of a number of actions and processes that help to strengthen us professionally and personally.

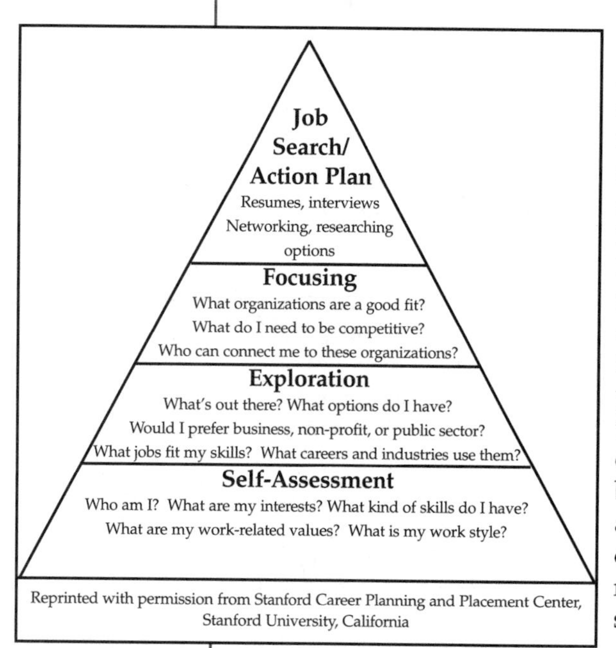

Reprinted with permission from Stanford Career Planning and Placement Center, Stanford University, California

Self-Assessment

As Chapter 5 will explain in more detail, self-assessment is the foundation of successful professional career development. Self-assessment is the process of evaluating one's own *skills*, *interests*, and *values*, the things we are best at and enjoy the most. Practically, self-assessment involves exercises, both self-guided and guided by a career counselor, that try (the operative word here) to assess what skills, values, and interests you have. While this is an inexact science, the process itself is valuable both because it enables you to ask difficult questions, and because it gives you clues about how you approach problem-solving and communicating with others and what other people—namely, colleagues, potential employers—see in you.

Exploration

Building from your self-assessment, exploration is the process of learning about the world of work. Chapter 6 of this book describes this in greater detail. In the broadest sense, exploration is the process of learning about different career fields and the personalities that inhabit them. One of the most useful and efficient ways for you to learn about different career fields is through informational interviews, which are extremely useful in building contacts and polishing the interpersonal skills you'll need during a job search. Exploration also involves identifying what skills you might need and how you can develop them.

One important area of employment you may think you already know is the world of research science in academia, industry, and government. In Chapter 7 you will find a summary and review of several excellent books that explore research science as a career.

Focusing

At the focusing stage, you narrow your search to specific fields that are a good match with your skills, interests, and values. Focusing involves identifying and contacting specific organizations and individuals for information about leads and openings. Typically, it is the focusing stage from which most Ph.D. graduates start when they explore the "traditional" fields of employment in academia, government, and industry. Chapter 8 explores this in more detail, and Chapter 7 provides some "focused" information about landing a job in research or teaching.

The 80:10:10 Rule: A True Story

Back when I was finishing grad school I was introduced to an alum of our department. After doing a postdoc, this person left research for a career in the oil industry, and had risen quite rapidly to a senior VP position in the organization.

I asked her what her secret was for success.

"The 80:10:10 rule," she replied.

I was not familiar with that rule, so she explained.

"80% of my work week I spend doing the best job I can possibly do. I really try to focus on the task at hand, I try to solve the important problems, and I really try to add value to my organization.

"10% of my work week I spend solely devoted to my own personal development. I close the door, turn off the phone, and explore issues and areas that I think might be important for my future. For example, last week I took half a day and met with two guys in our finance group and learned about ways of financing exploration costs. It was really cool!

"The final 10% of my time," she concluded,"I spend telling as many people as possible what a good job I'm doing ... the 80% of the time I'm actually doing my job!"

Job Search Strategy/Action Plan

The top level of the career planning pyramid consists of the actual activities associated with landing a job. These include writing an effective resume and cover letter and presenting yourself well in interviews. Chapters 9 through 12 describe this process in gory detail. As scientists we may believe that style should be subordinate to substance—our job materials should be judged on their intrinsic merit rather than their superficial presentation. However, in the case of most openings, employers are compelled to spend as little time as possible evaluating potential applicants. Thus, while substance is important, first impressions of a job applicant play a much larger role than they do in the academic or research job market. To overcome any preconceptions a potential employer may have about your capabilities because you have scientific training, it is critical to present your job materials in as professional and polished a manner as possible.

Career Planning Is Your Friend

Fundamentally, the process of career planning should be a natural for scientists because it involves gathering and assessing data, making informed and logical decisions, and aggressively pursuing new opportunities. However, when the data, decisions, and new opportunities involve our own livelihoods, the process can seem daunting. Career planning is like regular exercise: it takes a small daily commitment, it is usually enjoyable, and it can save your life.

What about head hunters, search firms, and placement agencies?

Many young scientists have asked me about the merits of utilizing the services of a placement agency or head hunter. It seems like an attractive solution, why sweat your job search yourself? Just get someone else to do it!

Unfortunately, a head hunter or placement agency may be less productive than you think. Placement agencies typically make their money by charging an up-front fee to you, the job seeker. Most placement agencies are, in fact, paid by corporations that are downsizing and are most accustomed to dealing with middle managers as clients. I know no agency that helps newly minted Ph.D.s and have not heard any success stories.

Head hunters typically charge a company a fee for finding a person. They are free to the job seeker. While this may seem like a good deal, there are some factors to consider. First, a head hunter works for the company—not for you, the job seeker. They will only take you on as a client if they think they can place you, and they will seek placement for you only from a company that will pay them. You may find your options restricted as a result. Second, there are only a few technical areas where head hunters are active. These tend to be hot technology areas such as information technology, computer science, and biotechnology. If you have the right set of skills you may be flooded with opportunities. In such a situation a head hunter might be of great value in helping you find the best opportunity and negotiating the best offer. Your head hunter would act more like your agent. But in most areas of science, head hunters are scarce to non-existent. Finally, consider this. Head hunters make the most money placing high-level (read: expensive) senior management positions. Not too many are interested in placing "entry-level" people—unless you are going to be making a LOT of money!

Temp agencies—which most people dismiss out of hand—may be a clever way of getting a foot in the door of a particular organization. Some of the larger temp agencies, such as Manpower and Kelly, have divisions that focus on technical professionals. If you are focusing your job hunting efforts on a particular firm, and that firm uses a temp agency to fill some technical positions, getting a position through the temp agency might be a good way of getting your foot in the door and exploring the organization. But this is a strategy that is best employed when you have FOCUSED down to a specific target, a stage of the career development process we will discuss at length in Chapter 8.

A Note About Career Books, Career Counselors, and Centers

Time and again, whenever someone finally decides they need to make a career change, the first thing they do is run out to the local bookstore and load up on career books. This may feel very satisfying—you have taken your first step and have made an important and valuable investment in your future. In fact, what you have done is just drop a non-trivial sum on a number of books that are likely to be hopelessly general, redundant, and from which you might glean only a few valuable kernels of wisdom.

Let's get one thing straight: there is a BIG INDUSTRY in publishing books and the purpose of that industry is to make as much money from you as possible! The quality and applicability of career books varies widely: the ones cited here are good, but others are poorly done and actually contain BAD ADVICE! Stop what you're about to do, hide the credit card, and just sit down and read the next two paragraphs.

If you are interested in sincerely exploring other career possibilities, the best place to go is not your local bookstore but to your local career planning and placement office. For most of you, your school or institution has a career center that you can use FOR FREE. Some institutions may not give postdocs access to their career planning and placement center. If you find yourself in this situation, contact your undergraduate or graduate institution and ask them to send a letter of reciprocity to the career center you want to use now. Most career centers will admit people given such a letter. However, if this does not work, a year-long membership to your community career planning and placement center is cheap, usually less than $50. And in some cases you can browse through most of the materials without a membership. Go to one, check out the environment, and look through the career books. After reading through some, you may find a few titles you like and think you will use regularly. Then go visit the bookstores!

The second piece of advice is quite personal: consider finding a career counselor you can talk to. This was the most valuable thing I did in my entire job search. A good career counselor will be extremely knowledgeable about the process of career planning and will have the necessary background in counseling. This person can really help you feel better about yourself and your options, much better than any book (including this one!). There are also really inept counselors out there, so be sure to "interview" the person first. If he or she has no experience with technical people you may want to look elsewhere. For those of you lucky enough to be in an institution that has free career counseling, you are simply crazy not to try it.

Summary

- Career planning is not a stop-and-start activity but a continual process of professional development.
- Successful professionals adopt good career planning habits throughout their careers.
- Career planning helps you identify your options and realize opportunities quicker.
- Career planning WORKS!

Now what?
AJL

Self-Assessment

Making Your Neuroses Work for You!

5

Make it thy business to know thyself, which is the most difficult lesson in the world.

Miguel de Cervantes

So you earned a Masters or a Ph.D. in some field of science and now you're wondering what on Earth you might do with yourself besides become a scientist. It seems like you've been doing science for as long as you can remember. And along the way, everybody told you that you were doing the right thing and to keep going. But now, for whatever reason, you are thinking about trying something else. Scary, isn't it?

Most people start by looking outward and asking questions like:

- *What careers or specific jobs use people like me?*
- *What have other grad students done besides research?*
- *What industries use my scientific skills?*

These are good questions, and it may seem tempting to get into specifics right away. But there is some important foundation work you have to do beforehand. You have to understand a bit about your own skills, interests, and values BEFORE you can decide what pathways might be a good match for you. So instead of asking questions like those above, better questions might be

- *Why am I looking for a change?*
- *What am I dissatisfied with in my current occupation?*
- *What do I really enjoy doing?*

Enter Self-Assessment

Nearly every career counselor (and quite a few other folks) will tell you that the process of career development or career change starts on a foundation of self-awareness. The basic philosophy is:

> If you love what you do for a living, you will probably be successful at it.

You would be surprised how many people like yourself begin their job search by focusing on specific career fields, and sending out scores of resumes rather than sitting down and figuring out what they love doing. Without some self-awareness about what you really want to do with your life and what really turns you on, you run the risk of proceeding down a path that will end up leaving you just as frustrated and stranded as you feel right now.

Self-assessment exists to help you identify your own skills, interests, and values. Armed with this information you will be better able to judge opportunities, chart new directions, and, hopefully, find a career that turns you on.

Self-Assessment Tools of the Trade

Self-assessment comes in many forms. There are self-guided exercises, role-playing games, written exercises, paired and group exercises, exercises with counselors, and other, stranger permutations. In fact, a vast part of counseling psychology deals just with this very issue.

Ideally, a sincere effort at self-assessment will leave you with a better understanding of:

- Who you are and what you like and dislike in work and play.
- How you work, play, and communicate with others.
- What your strongest skills are and where your weaknesses lie.
- What work environments you prefer.
- What your goals and personal values are.

You may also learn about career fields that appear to be a "good fit" with your skills, interests, and values.

Another important benefit of self-assessment is that it will help you communicate your skills, interests, and values to potential employers. Some of the exercises in this chapter will even help you structure responses to tough interview questions.

Formal Self-Assessment Exercises

Formal self-assessment usually is carried out with the assistance of a career counselor. Most exercises are in the form of a standardized inventory that you fill out, not unlike the Graduate Record Exam (GRE). Your answers are scored and you and your career counselor then take the results and discuss what directions look attractive. However, these tests cost money. The career planning and placement center at your college or university may provide such tests free or at a subsidized rate so you might want to check them out first.

The Myers-Briggs Type Indicator (MBTI)

There are several varieties of this most-famous-of-tests, but all are based on the psychological theories of Carl Jung, a Swiss psychologist and student of Sigmund Freud. Jung believed that people could be sorted into specific "psychological types." Carl Jung was the kind of guy who sorted all his Halloween candy before eating it. Actually, the four different variables that Jung proposed, and those that make up the MBTI, make some sense and are somewhat familiar. They are:

- Extroversion versus Introversion
- Sensing versus Intuition
- Thinking versus Feeling
- Judging versus Perceiving

From these four scales, people are sorted into one of 16 bins, such as INFP (Introverted, iNtuitive, Feeling, Perceiving). The book *Do What You Are* by Paul Tieger and Barbara Barron-Tieger describes the MBTI in full detail and profiles all 16 personality types and the types of careers they find fulfilling.

The MBTI is used a great deal in this country, not only for career counseling but by workplace psychologists and management consultants to diagnose dysfunctional relationships in the workplace. You probably ask yourself: "Why is my advisor such a meddlesome argumentative micromanager?" Because he is a jerk? No, it's because he's an ESTJ! Rather than feeling frustrated and bitter, your analysis of workplace personalities may help you adopt more effective strategies for dealing with the jerk... I mean, your advisor.

In the field of career counseling, the MBTI is used to help identify career fields in which you might feel more at home. For example, a person who scores high in Extroversion might be more happy in a job in sales or project management than a person who is an Introvert. A "Judger" might not make as good a counselor as someone who scores high in "Perceiving" and "Feeling."

The Strong Interest Inventory (SII)

The SII, like the MBTI, evaluates your responses to a series of questions about work situations, occupational likes and dislikes, and subject matter interests. The SII sorts personalities into two of six personality bins. However, the SII also takes your answers and compares them with answers from a host of control responses received from people in a variety of career fields who consider themselves "satisfied" with their careers. Thus, your answers are compared with those from a

> **Web Resource**
>
> **The Kiersey Temperament Sorter**
>
> (www.keirsey.com) is a free and easy personality type indicator that is available on the Web. Of course, once you fill out the survey you will be invited to purchase a book to explain the results but you may try checking it out of your local library or career center first!

group of broadcasters, hair stylists, college professors, city administrators, and others. This gives you and your career counselor two ways of understanding the results: on an absolute scale and by comparison with others. It can be interesting to see what careers most closely match your answers to the test. In my case, my answers were quite dissimilar to the career field of "geologist" but strongly similar to those from "speech pathologist," "college professor," and "executive housekeeper." Go figure!

The Career Beliefs Inventory (CBI)

Traditional career counseling and many self-assessment exercises try to match the skills, interests, and values of people with jobs. However, it is obvious that a number of other more complex factors come into play in seeking a career, such as our past experiences, mentoring, and the expectations of others. In many cases, what guides us toward some careers and away from others are beliefs we hold based on our past experiences. These beliefs may create barriers that prevent us from considering all of our options.

The CBI is designed to identify those barriers and to tease out answers to difficult questions like:

- How flexible are you?
- How much do you need to have certainty about your future?
- How much are you motivated by achievement and recognition?
- How competitive are you?
- How mobile are you?
- How much of a risk taker are you?

Like the other self-assessment instruments mentioned above, the CBI is a multiple choice test that is graded by computer. However, unlike the other assessment tools, the focus is on personal issues related to career choices.

One big concern for Masters and Ph.D. scientists considering a career change is the anxiety that their Ph.D. or Masters degree training will have been "wasted" if they decide to pursue a non-technical career. Even people who HATED their grad school experience still feel the compulsion to do something "useful" with their degree. As a result they end up focusing on technical careers even though they hated working in the lab. Issues of belief such as these are at least as important as skills-matching when it comes to finding a new career, and these beliefs may be creating enormous obstacles that you never knew were there.

Because the CBI is based on cognitive psychology it is particularly valuable for people who feel great anxiety and depression about their current career situation. Scientists who feel trapped by their training or pressured by their peers or mentors to consider only a few career options may benefit the most from the approach taken by the CBI.

Advice from the Field

The most common mistake that people make in their career decisions is to do something because they're "good at it." It's a story I hear all the time. Someone will say to me, "I'm an engineer, but I don't like it." Why did you become an engineer? "I was good at science and math, so people told me I should be an engineer." Did you ever like engineering? " "No, but it was easy."

James Waldroop—in *You Decide: Vocation* by Chris Rogers, Fast Company, February-March 1998

Skills and Values Card Sorts

Standardized assessment tools have a number of advantages. They are based on years of research, sound theory (sometimes), and provide a quantitative measure of important attributes. However, standardized assessments usually take several weeks to score, cost money—unless your school provides them for free—and often a career counselor is needed to interpret them.

Skills and values card exercises are a quick and easy way to explore the same issues by yourself. You take a stack of cards, usually listing particular generic work skills or values, and sort them. The sorting is done in a matrix. First, you sort the cards according to one axis, such as according to proficiency, from "highly proficient" to "little or no skill." Then you sort each sub-pile according to the degree to which you like using each of the listed skills, for example, from "strongly dislike using" to "totally delight in using." After constructing the matrix, you examine which skills or values you found desirable to use and in which you believe yourself to be proficient. Careers that depend on those skills might be a good match.

Skills & values card sort.

In contrast, skills that you do not have and/or hate using should be avoided. This seems rather elementary, perhaps even a bit pedantic, but you might be surprised how many people work in jobs that have little overlap with the skills they are good at and like to use.

The card sort process works well because it is quick, aiming to elicit a first impression about each listed skill. It is limited because it is a relative ranking based on self-perception and can be strongly affected by your mood from day to day. But this too can be a valuable and interesting exercise, especially if you keep track of the changes in a career change journal.

Lies, Damn Lies, and Career Tests

Lest you be left with the impression that career development tests are the answer to all your problems, let me qualify the preceeding section. Career development tests are not necessarily the best guide to a successful career!

There are a few books that purport to direct people to their "ideal careers" based on tests such as the MBTI. That sounds just a bit too reductionist to me. If you think about your present work situation you have to admit that more than one "personality type" might be successful in such an environment. In fact, one of the strengths of a team is that each member approaches things from a somewhat different perspective. The last thing one would want to have is a bunch of clones!

I have found that some of the most productive and innovative people in a profession have been misfits who did not fit the typical profile and personality of the organization. Sometimes it is the "dissonance" between a person and their profession that can lead to innovation and reform. These tests, except for the CBI, may indicate what fields MIGHT be a good match for you. But if your heart is set upon being a concert violinist and your test results suggest this is a poor match, GO WITH YOUR HEART.

Informal Self-Assessment Exercises

Informal self-assessment exercises are activities you can do and evaluate by yourself, in the privacy of your home or office and on your own time. While there are entire workbooks with scoring charts that you can use, some of the best informal self-assessment exercises are simple questions that you ask yourself. I strongly urge you to write down your answers because you'll focus on the question better that way. The important thing is to be as honest as you can. Here are some that I like from various sources.

Remember – it is important to actually sit down and write out your answers.

Dave's List

What I like/dislike about academia

Like	Dislike
1. Intellectual challenge	1. Grant rat race
2. Academic freedom	2. Pressure
3. Can wear socks and jeans to work	3. Lack of collegiality/ teamwork
4. Can set my own hours	4. Long hours
5. Smart colleagues	5. Obscure work (really benefits society???)
6. Interesting work	6. Having to move to find a job
7. Making a positive contribution to society	7. Lack of security (until tenure)
8. Teaching students	8. Narrow focus

1. Make a two-column list of everything you can think of that you like and dislike about a career in academia, and then assign priorities. What do you learn about your values, interests, and skills as they affect the work and workplace? (From *Outside the Ivory Tower* by Margaret Newhouse.)

2. If you could live five lives and explore a different talent, interest, or lifestyle in each, what would you be in each of them? (From *Wishcraft* by Barbara Sher.)

3. Think back over the experiences you have had in your life—in the areas of work, leisure, or learning—and pick five to seven that have the following characteristics:

- You were the chief or a significant player.
- You regard it as a success: you achieved, did, or created something with concrete results, or solved a problem, or gave something of yourself that you were proud of and are pleased with.
- You truly enjoyed yourself in the process.

List each of these events, write why you consider it a success, and write a paragraph or two detailing the experience, step by step. Extract from these stories the values and interests they reveal about you and the skills you used. In other words, what do they reveal about what you like to do and do well? (Adapted from *What Color is Your Parachute?* by Richard Bolles.)

Here is an example, done by a Ph.D. in marine sciences:

Director of Fundraising, Hawaii SPCA, 1998-2000

While pursuing a Ph.D. full-time at the University of Hawaii, I led an effort to raise $60,000 to fund renovation of the Hawaii SPCA Center in Honolulu. This project involved coordinating and organizing a group of volunteers, organizing a huge garage sale, and handling all the financing, accounting and publicity for a special event, the Hawaii Dog Run. It required good leadership and teamwork skills, the ability to motivate, organize, and execute many details simultaneously, hard work, and as I was unsupervised in my handling of all the money, trust-worthiness. It made me realize that I prefer working with people, working in groups, and taking a leadership role, though not the sole leadership role.

This exercise works well if you also tell your stories to one or two friends and ask them to reflect back to you the skills, interests, and values they perceive. This is also a good exercise from which to construct good interview answers and resume items.

4. Make a two-column list of "characteristics any job I take *must* have" and "characteristics it *must not* have." Making this list will help you summarize your knowledge to date and keep you focused on your central values and requirements. You can add to it as your career search progresses. It can also keep you from compromising on essential things when you get a job offer. (From *Outside the Ivory Tower* by Margaret Newhouse.)

Dave's List

Must have	Must not have
1. Intellectual challenge	1. Stupid bureaucracy
2. Relative independence	2. Boring/repetitive tasks
3. Casual atmosphere	3. Neckties
4. Pay above $50K/year	4. Lack of advancement
5. Smart colleagues	5. Socially non-redeeming work
6. Interesting work	6. Long commute
7. Within 20 minutes from home	7. Stupid people
8. Flexible hours	8. MS Windows
9. Daycare	

5. Look through the following list of work-related values, changing the terminology or concepts as they apply to you, and adding more general life values that you want to consider. Then rate the degree of importance that you would assign to each for yourself, using this scale:

1 = Not important at all
2 = Somewhat, but not very important
3 = Reasonably important
4 = Very important in my choice of career

Once you have ranked this list, copy down each value that you ranked as a "4" and rank those in order. Put this list in the inside front cover of your career change journal, where you will see it every day. As you go through your job search this handy list will remind you of the things that really matter in your career. (Adapted from *Outside the Ivory Tower* by Margaret Newhouse.)

Work-related values

________ *Social service:* Do something to contribute to the betterment of my community, country, society, and/or the world.

________ *Service:* Be involved in helping other people in a direct way, either individually or in small groups.

________ *People contact:* Have a lot of day-to-day contact with people—either clients or the public—and have close working relationships with a group; work collaboratively.

________ *Work alone:* Do projects by myself, without any significant amount of contact with others.

________ *Friendships:* Develop close personal friendships with people as a result of my personal work activities or have a career that permits time for close personal friendships outside of work.

________ *Competition:* Engage in activities that pit my abilities against others where there are clear win-and-lose outcomes.

________ *Job pressure/Fast pace:* Work in situations with high pressure to perform well and/or under time constraints; fast-paced environment.

________ *Power/Authority:* Have the power to decide course of action, policies, etc., and to control the work activities or affect the destinies of other people.

________ *Influence:* Be in a position to change attitudes or opinions of other people.

________ *Knowledge:* Engage in the pursuit of knowledge, truth, and understanding; work on the frontiers of knowledge, for example in basic research or cutting-edge technology.

________ *Expertise/Competence:* Being a pro, an authority, exercising special competence or talents in a field with or without recognition.

________ *Creativity:* Create new ideas, programs, organizations, forms of artistic expression, or anything else not following a previously developed format. (Specify type of creativity.)

________ *Aesthetics:* Be involved in studying or contributing to truth, beauty, and culture.

________ *Change or Variety:* Have work responsibilities that frequently change in content and setting; avoidance of routine.

________ *Job stability and/or Security:* Have predictable work routine over a long period and be assured of keeping job and a reasonable salary.

________ *Recognition/Prestige/Status:* Be recognized for the quality of work in some visible or public way; be accorded respect for work by friends, family, and community.

________ *Challenging problems:* Have challenging and significant problems to solve.

________ *Career advancement:* Have the opportunity to work hard and rapidly advance.

________ *Physical challenge:* Have a job that makes physical demands that I would find rewarding.

________ *Excitement/Adventure:* Experience a high degree of excitement in the course of my work; have work duties that involve frequent risk taking.

________ *Wealth or Profit:* Have a strong likelihood of accumulating large amounts of money or other material gain.

________ *Independence:* Be able to work/think/act largely in accordance with my own priorities.

________ *Moral fulfillment:* Feel that my work contributes significantly to, and is in accordance with, a set of moral standards important to me.

________ *Location:* Find a place to live that is conducive to my lifestyle and affords me the opportunity to do the things I enjoy most or provides a community where I can get involved.

________ *Self-realization/Enjoyment:* Do work that allows realizing the full potential of my talents and gives high personal satisfaction and enjoyment.

Self-Assessment Is IMPORTANT: Don't Skip It!

Self-assessment is not the answer to all your career woes, but it is a critical first step. However, these tests and exercises may reveal a few things that you had not considered, interests you did not realize you had, or values you didn't appreciate. The results might challenge your perceptions about what you think you SHOULD be pursuing as a career and may provide that bit of motivation to strike off and explore a new area.

Perhaps most importantly, self-assessment can help you clarify your goals and strengths and present them more compellingly to prospective employers. Understanding your own particular strengths and idiosyncrasies helps you develop pride in your individuality. This is what is known as self-confidence, and it comes through loud and clear come interview time.

Summary

- Self-assessment is a critical foundation to successful career planning. You can use formal or informal methods, with or without a career counselor.
- Information is power: Self-assessment brings with it a measure of understanding and control, not only for you in your job search but also for your interactions with other people.
- Self-assessment helps you prepare the best possible resume, cover letter, and interview responses.

Captain's Log: We've reached the black hole at the center of the Career Nebula, and the crew wants to fly into it just for fun. We'll probably die. . . but what the heck.
SS GRAD STUDENT
PHD-2001-SC13NC3

Beyond the Endless Frontier 6

Exploring the World of Work

When we are ready to make a beginning, we will shortly find an opportunity.

William Bridges
Transitions

To the average scientist, non-science careers in the "real world" may all seem a blur. We know the things that we do, but all that other stuff: law, business, politics, seems equally remote. Let's face it: the Ivory Tower happens to be a very apt metaphor for the world of research science in academia and many government laboratories. Unlike most other careers, in research science we are intentionally isolated from the concerns of everyday life. We grapple with the most profound of questions: the age of the universe, the origin of life and intelligence, the mysteries of tenure. As a result, we are more ignorant than most about how the world really runs. When it comes time to actually find gainful employment out there, this ignorance contributes to our fear and uncertainty.

How do scientists learn about potential career options and opportunities beyond science? By doing what they do best: research!

Learning about the world of work and the many opportunities that are out there requires you to lift your head up and look around. There are many sources of information out there for you to use, but they won't do you a bit of good if you don't take the time to explore them! Simply put: the more career information you acquire, the more opportunities will present themselves.

Sometimes the greatest opportunities are found furthest away from your present situation.

This latter concept may be a bit disconcerting to the average scientist who is used to a very linear and progressive career track, but I have seen it many times. Someone with a Ph.D. in geophysics may find his life's calling in running a successful bakery (this is a real person), or a Masters in chemistry may find it running around the Iraqi desert inspecting bombed-out nuclear weapons labs with the United Nations (also a real person). How about the Ph.D. biologist who has started her own science news service? (Yep, another real person!)

How Do You Start?

It is important to get into a routine in which you regularly investigate sources of information about potential careers. Think of yourself as an information filter feeder: the greater the flow of information past you, the more potential opportunities will come your way. If you have carried out some sincere self-assessment, as described in Chapter 5, you may have some indications of your own abilities and interests. Some self-assessment exercises may have pointed you in the direction of one or more specific career fields, probably some that you've barely heard of. Even without self-assessment you probably have at least a few questions about life and work in other careers. How do you find out if any of these careers would be appropriate for a scientist like yourself? Start by exploring these sources.

The News Media

Learning about the world of work requires that you know something about the world itself. Granted, you're not a complete troglodyte; you read the newspaper, listen to the news, etc., but you probably haven't done these things with the aim of learning about different careers. The newspapers are a great source of information about non-traditional careers. Are you a geoscientist? In newspaper and National Public Radio articles over the last few years it was reported (rather parenthetically) that the Rand Corporation hires geoscientists, as do the United Nations, Mitre Corporation, and the city of San Francisco. In each of these cases a bit more probing revealed details of what the geoscientists were doing in each of these organizations. Some of the careers sounded quite exciting. It also helps to read the relevant publications for the industries that interest you. Many of these are now free online.

Career Books

Aside from the chance discoveries of career information that you may find in the news media, career books and magazines are a great source of more in-depth information about specific career fields. There are a lot of books out there, on every conceivable subject of work. These books give a great overview to the general field of work and provide important information about the work environment and the general atmosphere. For example, there is an entire shelf of books about starting your own business. Many public libraries have good collections of career books, as do career planning and placement centers.

Laws of the Job Search

by David Maister

1. Ban the word "should" from your job search.
2. If your work doesn't turn you on, you won't be very good at it.
3. Changing jobs is easier than changing families, and a lot less painful.
4. The more confusion you feel, the worse the decision you'll make.
5. Remember, the point in life is to be happy. All other goals (money, fame, status, responsibility, achievement) are merely ways of making you happy, and worthless in themselves.

Career Books for Scientists

In 1996, when the first edition of this book was published, there was almost no information out there on non-traditional careers for scientists. Today there are some great sources of information about a wide variety of careers and the scientists who populate them.

On The Market (edited by Christina Boufis and Victoria Olsen) is a collection of stories and advice from the front lines of the academic and non-academic job market. Twenty-eight humanities and science Ph.D.s share their stories about the tough world of the academic job market and the possibilities that exist beyond the Ivory Tower.

The Guide to Nontraditional Careers in Science (by Karen Young Kreeger) is an excellent resource for science and engineering graduate students exploring non-research careers. Kreeger gleans advice from over 100 professionals in Science Education, Medical Illustration, Science Writing, Informatics, Technology Transfer, Business, Law, and Science Policy. Each chapter also contains lists of resources and pointers to more information.

Alternative Careers in Science (edited by Cynthia Robbins-Roth) presents the stories of 23 scientists who made career transitions to a variety of fields in business, government, and elsewhere. Robbins-Roth's contributors include a technical writer, broadcast journalist, venture capitalist, investment analyst, consultant, patent agent, corportate communications manager, policy maker, and head hunter. All of them come from a science background and all have great advice about finding one's own path.

Some professional and scientific societies also publish guides to careers in and out of science. The American Chemical Society, the American Physical Society, and the American Geophysical Union each have booklets on careers in their disciplines.

People You Know

People can be the best and most valuable source of information about various career fields. Not only do most people know a great deal about what they are doing, but they tend to know something about related career fields as well. They also know you! And since people generally like to talk about themselves and what they know, it is usually easy to learn about career fields from them. Rather than the formalized informational interview, which is discussed later, the more informal conversations with people are the most useful to get a "lay of the land." Have you ever talked to your uncle or aunt about what they do? I'll bet not. Have you ever talked to the parents of a friend about their careers? The people you know are great to

talk to about these things because they are your friends and can be trusted. Most importantly, you can relax and not fear sounding ignorant around them.

Your Local Career Center

Your local career center is a place that has an abundance of both printed material and personal knowledge. A career center usually has a good offering of books that are free to read (see Chapter 4 for my admonition about buying career books). Career centers organize lectures and panel discussions with people from various career fields. Some university career centers have a list of school alumni who have agreed to serve as contact points about various fields. You may be able to access your undergraduate institution's alumni database remotely. And then there are the counselors themselves, who are in the business of learning about the world of work and have some familiarity with many different career fields.

The Internet

Sure, the Internet is a useful information resource, but what can you find that is REALLY valuable in exploring different careers? Your first stop must be: Science's Next Wave—the career development site for scientists (www.nextwave.org). Next Wave has a huge range of resources and information about a wide variety of careers for scientists in and out of research and profiles of over 100 scientists who now work in non-science careers. Some professional and scientific societies are also expanding their career development offerings on the Web.

The Internet is, of course, far more than a source of career development information. It enables you to find information that was simply unavailable to job seekers as little as 5 years ago. You can now look up individual companies and obtain information on what they do and who they hire. You can find long-lost college friends who may be working in interesting fields. You can glean press releases, roam chat rooms, and connect with people anywhere in the world. And, of course, you can peruse millions of job ads.

Web Resource

Careers featured by Next Wave (www.nextwave.org)

Technology Transfer
Patent Law
Secondary School Teaching
Community College Teaching
Investment Analysis
Management Consulting
Entrepreneurship
Corporate Management
Technical Writing
Library and Information Science
Corporate Communications
Investor Relations
Science Journalism
Proteomics
Biomedical Engineering
Biomaterials Engineering
Forensic Science
Computer Modeling
Bioinformatics
Federal Science Policy
Manufacturing
Quality Control
Technical Services
Regulatory Affairs

Informational Interviewing

If you are seeking more in-depth, current and first-hand information about a career field, a particular company, or even a particular position, nothing will substitute for a good informational interview. An informational interview is a means of doing research and learning about a particular job, career, or organization. It is not about getting a job offer—at least, not directly. However, what Richard Bolles, author of *What Color is Your Parachute?*, and many other job placement experts claim is true: the contacts, information, and encouragement you generally get through an informational interview will very likely lead to a job opportunity.

What Is Informational Interviewing?

An informational interview is an interview you have with somebody wherein you ask specific and relevant questions about:

- Their job
- Their career and/or
- Their organization

If done correctly, you get in return information that is:

- First-hand
- Timely
- Accurate (from one person's perspective)
- Valuable

While there are many valuable aspects to the informational interviewing process, the biggest is that *it is by far the best way of getting information that will lead to a job.* I cannot emphasize this enough. It is rare that someone is offered a job right at the end of an informational interview, though it has happened. What is more likely to happen is that the person you meet will ask for your resume and pass it on to someone else who is seeking to fill a position. This is an important aspect of networking, a topic discussed below.

Career books and articles on informational interviewing list other advantages of the informational interview process as well:

- You are in control–you define the agenda with your questions. For people unpracticed in the art of the job interview, the informational interview can be a useful dress rehearsal.

- You can ask sticky questions that wouldn't be appropriate in a job interview. In a job interview, the role of the interviewer is very restricted: they are trying to get specific information from you and, at the same time, trying to make their organization seem as appealing as possible. In an informational interview, questions like "what do you like about your job?" are more likely to elicit a candid and honest response from your informational interviewee than from your job interviewer.

- You can see people in their actual work environment. In a job interview you are often one of several candidates, and you are usually whisked in and out as efficiently as possible. There is little time to be shown around the office. In fact, employers have to be scrupulously fair in their treatment of job candidates and this leads to a certain impersonal feel to the interview process. In an informational interview, no such rules apply. It is far more likely that you will get the opportunity to see some of the organization and meet other people, especially if you request to do so ahead of time.

- You can get feedback and advice. In an informational interview you can ask for and receive direct feedback about your resume, your interview style, and other aspects of your job search materials. In a job interview your only chance to find out what they thought about you is

AFTER you've been turned down. And there, too, the employer is under some strict rules about not showing bias in hiring. Believe me, it is much more comfortable discussing your resume and interview style in an informational interview than over the phone with the jerk who didn't give you the job!

How Do I Start Informational Interviewing?

Arranging for informational interviews requires a little leg-work. You really should have some referral, either a person, or a school alumnus/ae connection. Your network of friends and colleagues may be able to provide a contact. Often you will be surprised by who your friends know. Failing that, you may find a contact through the career planning and placement center you are using. Some career centers, especially those associated with colleges and universities, have lists of individuals who have agreed to be interviewed; alums commonly do. Once you get the first name, that person can give you advice about who else to talk to. Then ...

1. Contact the person by e-mail or phone, explain that you want to learn more about the career field and that you got his or her name from ____. Be clear about your intentions. If you are interested in getting a job in the organization, say so. If you are looking for more information about a specific career field, but not necessarily a job in their organization, make that clear as well. They may refuse or say that another person would be more appropriate. If so, contact that person and move forward.
2. Prepare some of your questions in advance and don't waste time. A typical informational interview lasts only 30 minutes. People do not enjoy answering questions that could or should have been investigated elsewhere. Be sure to ask for additional names of people you can speak to.
3. Questions asked usually pertain to:

The importance of the informational interview.

- Required background and training
- Specific information regarding the career
- Personal experiences
- Advice
- Future trends

Why Are These People Willing to Talk to Me?

The whole process might seem a bit strange to those of you who are unfamiliar with it. Why would busy professionals take time out of their day to discuss their job with me? Consider what is in it for the person granting the interview:

- Information transfer is a two-way street. Yes, you are asking them questions, but you'd be surprised what interviewers learn from the people who come through.

They could be interested in your research experience, information about other companies you have interviewed with, or any number of things. If they are alumnae of your school they may be eager to learn more about the recent goings-on at their beloved alma mater.

- People like talking about themselves. Not only is this true, but it works to your advantage. People end up telling a stranger (that is, you) things they might not tell their colleagues. Talk about an inside perspective! Furthermore, they'll like you more if you share their interests and values.

- People like to help. Many people believe that their success was due, in part, to the help and encouragement of others. Informational interviews are a way that they can return the favor. This is particularly true for school alumnae who used the "old boy" or "old girl" network to get their jobs.

- Organizations are always looking for fresh talent. Even though an informational interview is not a job interview, one of the questions that is in the back of the minds of the people you talk to is: "Would I hire this person?" Many companies spend a great deal of money on recruitment. The time it takes to talk to you is a small investment with a big potential pay-off.

Some Do's and Don'ts about Informational Interviewing

Informational interviewing, while still popular, has become a bit more difficult recently. I am hearing that some employers are tired of a plethora of requests for informational interviews from people who are not prepared or are just blatantly asking for a job rather than seeking job information.

In some ways, you should treat an informational interview just like a formal interview for a job. You should do your homework and read relevant background material, especially that which is available on the Web. You should think carefully about what you want to learn. It is a good idea to list your goals in your career change journal before you go for the interview. Preparing some questions in advance is also a wise idea. During the interview you should ask questions with the same professionalism as you would during a job interview; even though it is not a formal job interview you ARE, to some extent, being evaluated. Finally, you should always, ALWAYS write a thank-you note.

A thank-you note is warranted after an informational interview because the person you visited took time out of his or her day to help you. It is a gracious, classy, and professional way to demonstrate your gratitude. Thank-you notes appear to be a lost act of civility these days and many people fail to write them at all. However, consider this: if you are one of those people who DOES write a nice thank-you note you will stand out from the rest. You will show yourself to be a thoughtful, organized, and considerate person—just the kind of person that people want to work with. It is a demonstration of the kind of professionalism and thoroughness that wins clients, for example. This is not a bad final impression to leave with someone you have spoken to. Plus, you will be sure that they have your contact information.

In other ways, however, it is important to treat an informational interview differently from a job interview. Your goal for an informational interview is to gather information and build your network—not ask for a job. Keep this clear to yourself and to those you interview. Also, realize that what you learn from a single person may not be representitive of the views of the company or organization. People have their own perspectives and agendas. However, speaking with one person is still orders of magnitude more valuable than speaking with nobody at all!

Conducting an informational interview by phone is a good option if you are pressed for time or are too far away to interview in person. While you may lose some immediacy and the opportunity to check out the work environment, a phone conversation is easy to set up and very time efficient. Just be sure you prepare as thoroughly as if you were meeting in person!

Your Assignment

Set up an informational interview with someone you know—a friend who you feel comfortable with and who can give you constructive feedback. This trial run will help you work the bugs out and will make you more comfortable with the whole process. Don't put this off, do it soon. You'll be glad you did.

Networking: How Most of the People Around You Got Their Jobs

Many young scientists I have spoken with know the term "networking" but assume that it involves some smarmy, manipulative, and socially aggressive means of forcing yourself on people. Some people get discouraged, thinking that only someone with the skills of a used car salesman has any hope of successfully networking.

Nothing could be further from the truth! Networking is not some hypersocial, artificial activity that you must engage in to get a job. Rather, it is the process of meeting people and:

- Learning about careers and specific job opportunities from them
- Alerting them to your career goals and abilities

Networking is simply the process of doing research about the job market by talking to people. There is really nothing more to it than that. As a scientist, you can appreciate the need to conduct background research when you encounter a new area of science.

To borrow an analogy from nuclear physics, think of networking as increasing your job "capture cross-section." If you are alone in your job search, the only opportunities you will encounter are those that you find yourself. If you have the help of others, the probability of capture increases. A very large network can provide you with more leads than you can imagine. Wouldn't that be nice?

When you think about it, we scientists are not as unfamiliar with networking as we may think. After all, we go to scientific conferences where we meet fellow scientists and find out about their research. We talk to each other about potential postdoc and employment opportunities and a range of other issues of professional concern. Some of the best scientific connections have been made through networking. In the outside world it is very similar to the networking you do as a scientist. Not only do you use the same tools, you also ask very similar questions. Below is a table of activities you have probably participated in as a scientist and their job hunting equivalents:

Science Activity	Job Hunting Activity	Similarities
Poster Session	Job Fair	Organized event Presenters are organized into booths One-on-one contact with experts Two-way flow of information Reprints/company brochures available
Seminar	Recruiting Presentation	Lecture format Focus on one topic Q&A afterwards
Lab Visit	Informational Interview	Initiated by information seeker Site visit One-on-one contact with "experts" In-depth questions about topic Expression of interest in topic
Literature Search	Research on Company	Use of library, Web and other information sources Self-directed Goal is to provide thorough background on topic

Networking is a CRITICAL aspect of professional development in ANY career, especially research science. Networking enormously increases your chances of landing a good job and having better job mobility.

Who Is My Network?

Your network is anyone to whom you are willing to talk about your job search and your job questions and is someone with whom you are able to keep in touch. Ideally, it will be made up of people who care enough about you to actually keep their eyes and ears open for opportunities that might suit you. Here are a few specific suggestions:

- Fellow students, colleagues, and coworkers
- Relatives
- Past employers
- Scientists you meet at seminars, conferences, and workshops
- Neighbors
- Alums from your school
- Other people looking for jobs
- And people they know...

In other words, ANYBODY to whom you feel comfortable talking, even people you don't know.

How Do I Network?

Networking can be as formal or informal as you like. Some people keep a journal and make detailed notes about each contact. I find it easier just to make a point of asking people about their jobs when I meet them, like at a party. If what they do sounds interesting, I may delve further, and if they are doing something that REALLY sounds cool, I will ask if I can call them and find out more.

Informational interviewing is different from networking because it is a relatively formal process. Typically, it is carried out during a visit to someone's place of work. In contrast, networking is done anytime and anyplace and involves casual contacts. Rather than seeking specific information, it mostly involves keeping your ear to the ground for opportunities.

Networking on the Net

All sorts of people use the Internet. If you are a regular poster to a newsgroup your thoughtful words of wisdom may be read by thousands of people. You'll never know when someone might be very impressed with what you say. Often, private correspondences result. This, too, is networking. However, because networking on the Net is text-based rather than person-to-person, one has to be extremely careful in how one communicates. Sometimes people can be offended by casual or sarcastic remarks that you might intend as merely humorous. Because you lack any of the direct visual cues you get from speaking to someone in person you can easily blunder. Also, poor spelling, bad grammar, and other mistakes fail to impress anyone. Be careful about what you post to a newsgroup or on the Web. If it looks sloppy, poorly thought out, or careless you are not doing yourself any favors. You never know who is lurking out there.

Exploring for Life

Exploring for new opportunities, in work and in life, is more than a good career development habit: it is a characteristic of successful individuals. It is all too easy to become so busy that you give up looking for new opportunities—especially when you are young and immersed in graduate school. However, the marginal time you save

How your network works for you

A true story

The phone rang.

"Hello, I'm ______ and I'm a screenwriter in LA working on a script for a disaster movie about an earthquake that devastates San Francisco. Your cousin is a friend of mine and he said that you have a Ph.D. in geology and that you know a lot about this very issue and that I should talk to you."

"Well, yes," I told her. "I'd be happy to help."

"Well, what I really need is someone to read over this draft script and correct any gross scientific inaccuracies and provide some realistic-sounding disaster preparedness organizations. The script is 300 pages."

"Sure," I said. "Send it up."

"Gee, that's great. Of course we would pay you as a consultant."

"Well, that's not really necessary, but, uh, what does a scientific consultant earn in Hollywood?" I asked.

"Is $100.00 an hour OK?"

Pause.

"SEND ME THE SCRIPT!" I said.

by skipping exploration now will result in an extremely costly and time-consuming process of exploration sometime in the future. Developing a regular habit of looking for new opportunities is the best way of ensuring that golden opportunities will be available in the future ... when you will really need them!

Summary

- Exploration is simply keeping your eyes and ears open to possible career opportunities.
- Informational interviewing and networking are by far the most effective means of finding a new job and a new career.
- Keeping connected through networking is part of building and maintaining a professional career in any field, including research science.

The Ivory Tower Syndrome.

Exploring a Career That You Know

Research Science

7

Many professional scientists believe that "good" students find their way on their own, while the remainder cannot be helped. This justifies neglect, and perhaps not incidentally, reduces the work load. There may be some sense to the Darwinian selection process implicit in "benign neglect," but on the whole, failing to teach science survival skills results in wasting a great deal of student talent and time, and not infrequently makes a mess of students' lives.

Peter Feibelman
A Ph.D. is NOT Enough!

For most graduate students and postdocs, research science represents the "traditional," "intended" career path. While we may now be considering a wide range of "non-traditional" career fields, it was our interest in science that originally drew us in. Many young scientists wrestle with the question of whether or not to pursue a "traditional" career in science or to pursue a "non-traditional" career field for several years. However, few take a step back and consider the career path in science objectively. It may be a perfect match for your skills, interests, or abilities, or maybe not. But until you to give the career of science the same amount of consideration and scrutiny that you might give to an "alternative" career you won't know.

Getting a job and maintaining a career as a scientist requires all the same strategies and techniques that are useful for developing a career and finding a job in the outside world. As scientists, we pride ourselves on the meritocracy and fairness with which we conduct our business. However, all but the most naive of us would admit that getting a job in academia depends on a number of more, shall we say, subjective factors. Your advisors and mentors have told you, "if you're the best, you'll get a job." These days, not only must you be the best scientist, but you must be many other things: an effective communicator, a persistent fund-raiser, and an organized and efficient manager.

Not All Science Careers Are Equal

As a young scientist you have spent most of your professional life in academia. While it may have its pluses and minuses, it is certainly the environment with which you are most familiar. Because of this familiarity, it is very natural to gravitate to a career in academia. Your advisors and academic mentors probably think that academia is a great place to wind up—after all, they are there!

However, academia is but one of a number of potential pathways in a scientific career. Most graduate students receive little or no exposure to life outside the ivory tower—in government labs, industry, or non-profit research centers. Some advisors deride these other science careers as being inferior in intellectual rigor or prestige. As a result, your unfamiliarity can lead you to dismiss some potentially exciting and rewarding career opportunities.

While many of the same skills are required for a successful career in academia and other places of science, such as industry or national labs, the interests and values supported in these organizations may differ substantially. As we discussed in Chapter 5, skills, interests and values are all critical considerations in choosing a career. Depending on who you are and what you like, one of these settings may be much more satisfying and rewarding for you than the others.

The Life Cycle of a Scientist

We are all familiar with the traditional path of a science career: Undergrad-Grad School- Postdoc-Postdoc-Postdoc (OK, I'm just kidding here!) - Assistant Professor-Tenured Professor- Retirement-Death. In fact, the certainty of this process is probably one of the things that might have drawn you to a career in science in the first place.

In reality, this traditional career path is not followed by most scientists. Less than half the Ph.D.s trained in the United States end up in academia. Those that do end up in the ivy-covered halls of academia often take a more circuitous path – with stints in industry or a national lab. These are not detours—they represent the natural evolution of a research career.

Unless you have a very wise mentor, or a collection of advisors with experience in academia, industry, and other research environments it is likely that you will be missing some important information about your options and opportunities in a science career. Fortunately there are some terrific books and other resources that discuss some of the life cycle steps and career options open to scientists.

Grad School and Beyond

For those of you who are thinking about, or have recently begun, a graduate program, there are several valuable and entertaining books you should examine. Each covers a slightly different portion of the life cycle of a young scientist and all contain valuable information that will likely be immediately applicable in your life.

Getting What You Came For: The Smart Student's Guide to Earning a Masters or a Ph.D., by Robert L. Peters, Ph.D., is an informative and enjoyable discussion about how to survive and thrive in graduate school and beyond. Most of the advice is directed toward Ph.D. and research oriented students, but the more general advice about admissions and survival tips for graduate students is applicable for both Masters and Ph.D. students.

> ## Web Resource
>
> **Careers in Science and Engineering**
>
> **A Student Planning Guide to Grad School and Beyond...**
>
> Published by the National Academy Press, *Careers in Science and Engineering* is a concise guide for graduate students and post-docs planning their careers. It contains profiles, scenarios, and valuable advice about key skills that are needed for success in today's science job market, and practical advice about finding a job that fits. Best of all, the entire contents are available on the Web at no charge!
>
> (http://bob.nap.edu/html/careers)

The first six chapters of Peters' book address issues for the undergraduate contemplating graduate school. First and foremost is the question: "Do you need to go?" This question is not asked often enough, if at all, of undergraduates by their undergraduate professors, their family, or by themselves. The next five chapters focus on the application process and a practical description of what a Masters and a Ph.D. entail. The information in these chapters is very basic but extremely important and will probably be of greatest value to undergraduates and first-year grad students who have not had much exposure to life in graduate school. Peters devotes five chapters to the thesis. Interspersed throughout are some funny and terrifying anecdotes about theses, committees, and advisors.

All the remaining chapters, save one, deal with life in graduate school. One section deserves particular attention: "Dealing with Stress and Depression." Depression is one of the most serious and dangerous occupational hazards of graduate school. Graduate school puts enormous strain on all people, students *and* faculty, and even the strongest can be reduced to rubble. I am not aware of any other practical guide to college or graduate school that deals with this subject with such candor. Stress and depression are not signs of weakness, just humanity! Peters makes a great contribution in bringing this subject the attention it deserves.

The Ph.D. Process, by Dale Bloom, Jonathan Karp, and Nicholas Cohen, is the most comprehensive guide to date about graduate school in the sciences. Like *Getting What You Came For, The Ph.D. Process* marches through all aspects of the grad school experience from application through dissertation defense and includes some excellent topics not covered in other books, such as advice for foreign students. But unlike Peters' book, *The Ph.D. Process* is more specifically focused on the sciences. In addition, the book includes the voices of graduate students themselves, discussing, and in some cases qualifying, the authors' advice. The combination of authoritative summaries along with anecdotes from students themselves helps lend substance to what otherwise might be a daunting litany of do's and don'ts about grad school.

A Ph.D. is NOT Enough!—A Guide to Survival in Science, by Peter J. Feibelman, is a concise and witty survival guide for young scientists who aspire to a career in research. Feibelman discusses all aspects of a young scientist's career, from graduate school to succeeding in a science career. Feibelman wisely emphasizes the critical role of mentors, research "aunts and uncles," and how they can give a young scientist a broader perspective as well as access to the inner workings of the scientific process. Feibelman's

A Ph.D. is NOT Enough: The Cliff's Notes

Choosing the right advisor and thesis

- Choose established advisors with track records of success; they are more able to support you and they won't be competing with you once you finish.
- Choose a research group that is active, collaborative, and interdisciplinary; make sure the advisor and the members of the research group see the "big picture."

Choosing the right postdoc

- Choose a project and group that will enable you to finish up and publish research in time.
- Choose a senior scientist with an established research group.
- Don't be a slave.

Giving talks

- Never overestimate your audience: some basics are good because they let the audience feels like they know something.
- Make it clear what the big picture is.
- Make yourself heard and understood; don't talk fast.
- Cut the filler (outline slides, detailed experimental set-up diagrams).
- Make your overheads readable but not ostentatious.

Publishing papers

- Plan your research as a series of short complete projects: these are easier to write up and they keep your name in circulation.
- Write compelling papers (read the book for some very detailed advice here).
- Don't be afraid to use first person/avoid overuse of third person.
- Send it to a colleague for review first.

Choosing a career path

- Weigh the relative value of prestige, money, and security.
- Weigh the desire of teaching with the desire of doing research.
- Weigh the relative value of an academic setting against other environments.

Job interviews

- Do your homework; find out the issues and desires of the department or research group.
- Think of answers to obvious questions in advance: e.g. "What will you do here if we hire you?"
- Practice your %^$#!! talk, practice again, and again.

Getting funded

- Start writing grants in grad school, attend workshops if available.
- Address important issues, not just topics in your sub-discipline.
- Do not promise the moon.
- Get involved with a group effort.

Establishing a research program

- Be problem-oriented, rather than a technique-oriented thinker.
- Plan a series of short projects that drive toward a bigger goal.
- Remember that ambition is rewarded.

book is essentially a pocket mentor for those who might not have a strong one of their own!

In a just universe, all Ph.D. advisors would be able to dispense the sort of advice that is contained in these three books. Why don't they??? One reason may lie in the nature of the Ph.D. training itself. Many advisors adhere to the philosophy that survival instincts in science can't be taught and that only those with the right instinctive skills should survive. As Peter Feibelman states in his book, this notion is not only absurd but often ends up making life for students much more difficult than it has to be. Most often, the students who seem to be "on-the-ball" are those who had the fortune of receiving good career advice, either from a professor, more senior graduate students, or from a family member early on in their careers. It is possible to survive and thrive in graduate school and beyond. These three books may help you do it.

Beyond Graduate School: The Postdoc

A generation ago, a postdoctoral appointment was unusual. Most often, a bright young Ph.D. would proceed directly to an assistant professorship or a staff position in a research organization. Today, the postdoc is such a common part of young scientists' training that many institutions now call people in them "postdoctoral *students*." For some fields, such as engineering, postdocs are rare. In other disciplines, such as the life sciences, one can expect to remain in a postdoctoral position for nearly as long as one was a Ph.D. student.

For better or for worse, a postdoc has become a necessary step. The most widely cited reason is to gain more research experience, broaden your perspective, and build your record of publications and grants. Quite a few young scientists also use a postdoc to shift from one discipline to another.

I have met a few people who have chosen specific postdocs for more unusual reasons. One person deliberately targeted her postdoc search to facilitate her departure from research. Wanting to switch from research to science policy, she found a postdoc in the Washington, D.C. area. Knowing that proximity is enormously important in building a network, she stayed at her research postdoc for 2 years, making friends and professional contacts the entire time. At the end of her postdoc she made a smooth transition to a position at the National Academy of Sciences.

Some postdocs can give new Ph.D.s practical industry experience. Such postdocs do not necessarily have to be in industry: many universities and national labs have projects that are jointly supported by industry. Postdocs associated with these projects can literally work in both environments.

Some postdocs can provide a critical "foot in the door" to specific

Web Resource

Advisor, Teacher, Role Model, Friend:

On Being a Mentor to Students in Science and Engineering

The mentoring relationship between advisor and advisee is one of the foundation elements of the scientific education. However few advisors receive any formal training in how to be an effective mentor. This book, published by the National Academy Press, is a concise guide for advisors and their students. It contains helpful advice for mentors, case studies, and a host of resources and references. Plus, it's freely available on the Web.
(http://stills.nap.edu/html/mentor)

organizations. National laboratories, for example, hire most of their permanent staff from their ranks of postdocs. For the labs, the postdoc appointment is a valuable "test-drive" of potential colleagues. For postdocs, it is an opportunity to determine whether the national lab environment is the right place for them.

Postdocs in academia can provide young scientists with the opportunity to develop real teaching experience. The competition for academic teaching positions is so fierce now that teaching assistant experience alone is not sufficient for many schools. Real teaching experience, including course design, is essential.

These various motivations, though different, are ALL legitimate reasons for considering a postdoctoral position. The only poor reason to consider a postdoc is: "I don't have a clue what else to do."

Ingredients for a "GOOD" Postdoc

Despite the wide variety of motivations that young scientists have in choosing a postdoc, I have seen some common attributes and themes in the postdoc experiences that have been "good" or "successful" and those that have not. Most of these attributes can be ascertained by simply researching the position, talking to people, and asking some questions.

Good Attribute #1

A flexible advisor/project. This seems to be the single commonality in the many good postdoc experiences I have heard about. People in these situations have been blessed with advisors who have given them ample freedom to conduct their research and develop new projects. Many of these allow the postdoc to develop financial independence (see #3) by obtaining grants. These situations are superior because they best approximate the real world of research science, but they retain some important safeguards, such as a hard-money salary and supportive senior colleagues.

Good Attribute #2

Decent salary/benefits. The better postdocs not only have the attributes noted above, but they also tend to pay better. It seems that, in many institutions, increased compensation comes with increased intellectual independence. In institutions such as the national labs, postdoc salaries can reach as high as $85,000 a year, and postdocs tend to be treated like peers rather than lab animals. In contrast, the poorest paid postdocs, most often in universities, seem to be chained to a single project or piece of equipment in a capacity that is little more than a glorified research assistantship.

Good Attribute #3

Financial independence. Some postdocs are not tied to a single project, or even to a single advisor, but are free to develop their own collaborative research projects with whomever they please. As stated earlier, these opportunities tend to be better because they expose young scientists to the real world of research.

Good Attribute #4

Many good colleagues. The number and quality of peers and senior investigators, and the level of interaction between them, is another hallmark of the best postdoc opportunities (the same goes for graduate departments). The larger and more open the community, the richer your intellectual experience will be and the greater the opportunity to build new collaborative projects.

Good Attribute #5

Good facilities. While good facilities are a necessary condition for a fulfilling postdoc, the opposite—limited or bad facilities—is a major liability. The years spent as a postdoc are a critical time in the life cycle of a scientist, and can be completely wasted nursing poor or obsolete equipment. A parallel consideration is the cost and accessibility of the equipment that is functioning. Some organizations, such as the national labs, recoup high overhead costs by charging exorbitant fees for analytical equipment. If you lack a hefty budget, expensive equipment is no better than no equipment at all.

Good Attribute #6

Prestige. Prestige postdocs are those that are nationally recognized, heavily sought after, and have an illustrious track record of alumni. Getting one of these plum positions is a real feather in one's cap. But, like any job, its value to you is what you are able to make of it. If a prestige postdoc lacks the traits mentioned above it may be very difficult for you to live up to the legend, no matter how prestigious the position.

Oh, The Places You'll Go: Considering Academia, Industry, and Government Labs

Academia

Even though you've spent most of your career in academia, it has most likely been in only a narrow range of institutions. You may have gone to a small liberal arts college or a big state school as an undergraduate, but in graduate school there is a 90% chance that you are in a R-1 university. What's that? R-1 is one category in the college and university classification scheme put together by the Carnegie Foundation. R-1 universities award 50 or more doctorate degrees per year and have at least $40 million in annual federal research support. And these universities produce most of the Ph.D.s in science. Other types of colleges and universities have different environments, priorities, and faculty needs. It is critical to understand the differences as you consider your options. Several books are now available with excellent advice about the academic job hunt.

Tomorrow's Professor: Preparing for Academic Careers in Science and Engineering, by Richard M. Reis, is without a doubt the best resource for those considering a career in academia. *Tomorrow's Professor* is a comprehensive guide to academia, with exhaustive information about the job hunting process and valuable insider tips. Reis, a professor at Stanford University, provides an excellent insider's perspective on how candidates are critically evaluated for faculty jobs. His insights into how to be successful in the early years of an academic career are right on, and the book also contains a large appendix with examples of job application materials.

On The Market, edited by Christina Boufis and Victoria C. Olsen, is a collection of essays about the academic job search, written by and for aspiring academicians. Rather than simply cataloging helpful tips and advice, the editors wanted to create "an emotional guidebook to the academic job search," and by doing so create a resource that is part career guide and part therapy session. Those of you who are not presently engaged in a job search may not appreciate the value of the dialog. Those who are searching for their dream job are often in isolation and far from the support and camaraderie of their grad school compatriots.

Most of the writers in *On the Market* have Ph.D.s in the humanities (only four have Ph.D.s in the natural sciences or mathematics), and nearly half hold Ph.D.s in English. Because of this, science Ph.D.s might think initially that this book is less applicable to them. In fact, the academic job search, especially for smaller schools, is very similar in the humanities and the sciences, and we scientists would be wise to learn a bit more from the tribulations of our humanities brethren. In many ways, we in the sciences are much better off. The Modern Language Association (MLA) holds its annual meeting between Christmas and New Year's, guaranteeing that job seekers will have a perfectly miserable holiday.

Another important resource for the aspiring academic is The *Chronicle of Higher Education. The Chronicle* is THE trade publication for colleges and universities and features articles about higher education issues, profiles of movers and shakers in the world of academia, job advertisements, and other valuable information. In recent years, *The Chronicle* has published a number of articles about the academic job market and the changing scene in higher education. While most academic job openings in science departments are advertised in specific trade and professional publications, from time to time there are ads in *The Chronicle* that appear nowhere else. Anyone seriously considering a career in academia, either as a professor or as an administrator, should familiarize themselves with this publication. *The Chronicle On-Line* (http://chronicle.com) contains weekly stories, job ads, and links to other Internet resources.

The American Association for the Advancement of Science has developed an excellent site for young scientists:

> *The Career Development Center for Postdocs and Junior Faculty*
> (http://nextwave.sciencemag.org/feature/careercenter.shtml).

This site has regular features such as GrantDoctor, a clinic where readers can post questions about funding strategies. The site also has discussions on professional issues such as setting up a laboratory and manaaging graduate students. The site is also linked to GrantsNet, the largest and most comprehensive database of grant information in the biomedical sciences.

Industry and National Laboratories: There IS a Difference!

Most graduate students (and faculty) tend to lump jobs in both industry and national labs into a single category: places where you can do science but you have no freedom, are at the whim of bureaucrats and business people, and you can't publish what you work on. This common misconception is wrong in nearly all aspects. Industry and national lab jobs are not only better than most academics believe, but they are also very different from each other. Most importantly, it is industry and the government who employ most Ph.D. scientists! Dismissing job opportunities in these sectors cuts off a majority of the jobs that are out there.

Before I continue I should, in the interests of full disclosure, tell you a bit about my background. I am presently a staff scientist at a Department of Energy laboratory: Lawrence Livermore National Laboratory, in California. I have friends in equivalent research centers run by NASA, NOAA, the USGS, and NIH and who are research scientists in industry, in oil companies, biotech firms, and software development. So I come at this question with quite a bit of personal experience, both direct and indirect.

Broadly speaking, scientists in industry and national labs do the same work as those in academia: they use science and technology to answer critical questions. In industry, those critical questions are usually related to the creation of a product or a service that the company can sell. In the case of a national lab, those critical questions are defined by the mission of the sponsoring agency. To a large extent, scientists in academia are under the same constraints. Rather than answering to product managers (in industry) or program managers (in national labs), scientists in academia answer to funding agencies!

Scientists in industry and national labs tend to work in research groups similar in size to those in academia. A senior researcher may head a group of more junior research staff members along with a few postdocs. However, unlike in academia, group leaders in industry or at a national lab DO answer to someone! In industry, it may be the product manager or the chief technology officer (CTO). In a national lab, a PI would answer to a division leader who, in turn, might answer to a directorate leader or a lab director.

While the company or the sponsoring agency may define the research priorities in industry or in a national lab, scientists have great freedom in how they approach the subject of study. In this regard, industry and national labs tend to be less risk-averse than academia, and risk takers who succeed tend to be well-rewarded. PIs in these arenas are also more likely to sub-contract out research to other labs or to academia.

Industry and national labs differ significantly on academic-like issues such as the role of publications and funding. A strong publication record IS important to the career advancement of a scientist at a national lab. Publications in the prestige journals help the lab bolster its scientific credibility. Overall, however, the publication productivity of scientists at national labs tends to be lower than in academia. One big reason: scientists in national labs don't have graduate students to do their work for them!

In industry, publications may or may not be relevant. In many biotech companies, scientists are rewarded for maintaining a strong publication record because it brings attention to the company and helps the company recruit

other scientists. In more mature industries, publications may matter less. Some industries organize their own conferences, and presenting papers there may be encouraged. Sometimes, industry scientists will go to meetings just to search out the latest hot discovery that they may be able to use in the development of a new product.

One of the nicest aspects of working in industry, or at a national lab, is that it is often much easier to get financial support for your work. If you are doing research that is well-aligned with the needs of the organization, funding is usually no problem. Obtaining money for such research may involve nothing more than a memo to the CTO or program leader! Scientists with good ideas and the ability to communicate them tend to have no problem getting the resources they need to do their work.

But is intellectual freedom compromised? Are scientists in industry and national laboratories less able to pursue their own research? In some senses, yes! Companies have products and national labs have missions. If your research isn't addressing the needs of your employer, you don't end up contributing much to the organization. And that is not a great recipe for advancement.

That being said, it is only fair to compare academia, industry, and national labs on an equal footing. After all, you're not exactly free to do any sort of work you want in academia either—you have to get funding! If your work doesn't fit into the scope of the granting agency or program, you are unlikely to have much success ... unless you already have tenure! No matter where you work, your research has to be important to SOMEONE. If it isn't, you should be asking yourself why you're doing it in the first place.

Industry and national labs differ from academia significantly in terms of lifestyle and atmosphere. If you've ever paid close attention to the life of your advisors you probably know that they have some freedom: they are free to work any 80 hours a week they want! In all honesty, people do not have to work as many hours as their academic kin. Some in academia are a bit judgmental about this fact, thinking that industry and national lab scientists are "lightweights." Maybe so. But those "lightweights" manage to see their families, enjoy raising their children, and in general carry out those activities we have come to label HAVING A LIFE. Everyone is different in how they may prioritize work and play. For some, the flexibility of a regular work week is a real asset. However, others—especially young hard-chargers fresh out of school—can find the slower pace frustrating.

Compensation and benefits are another issue. Yes, they differ somewhat. In general, industry salaries start off lower than those in national labs (except at NIH, which has some of the lowest salaries in science), but they can rise far higher as one advances, especially if one takes on management duties. For labs run by the federal government, there is a real problem with salary advancement, especially for senior scientists. I am fortunate to work for a national laboratory that is run by an outside contractor, the University of California.

Job security is also a major difference. At national labs, it is very rare that staff members are laid off—for any reason! As a result, there is some fraction of the staff that really does not do much. It turns out that industry tolerates quite a bit of dead wood too, when times are good. But when the industry sector or the economy as a whole turns sour, entire R&D groups can be dismissed. Often, those who are displaced need to retool or find jobs in other firms, and this can be a very traumatic process. Imagine going through all the difficulty you're facing presently in your job search but add on a mortgage and two kids in college. Yikes.

How do I choose?

For a young scientist, choosing whether to work in academia, in industry, or in a national lab comes down to weighing what is most important to you. Some fields of science are only well-represented in one of the three sectors, making your choice easy. Others with a choice might weigh issues such as security, intellectual independence, work life, and compensation. Remember: one size does not fit all. Many professors in academia are terribly judgmental about work in industry or national labs. Don't let their choice automatically become yours, too!

Do I stay or do I go?

Many young scientists struggle with the question of whether, or when, to "give up" on a research career and pursue something else. Often this question comes down to opportunity costs. Opportunity cost is the economic term that describes the cost of a choice in relation to all the other choices that one could have made. By choosing to remain in science, for example, by taking another postdoc, you keep alive your chances of eventually finding a permanent job. However, at the same time, you forgo potential opportunities to move in a different direction. It is a truly agonizing decision. One postdoc described the situation perfectly:

> "I am piloting a plane, and this plane is running out of gas...I am trying to find a safe place to land before I crash. How long do I try, before I 'punch out' and hit the eject button?...Now add my [wife and two children] to the plane as passengers. Should I eject with them earlier than I otherwise would?"

Summary

- A career in science requires planning and career development just like any other professional career.
- Research jobs in academia, industry, and national laboratories are all very different. One of these might be perfect for you.
- A career in science can be your primary goal, but it does not have to be your only goal.

You can't decide whether to go into regenerative microsurgery or the Imperial brute squad? I think we can narrow this down.
AJL

Focusing on Specific Opportunities

8

Going directly to places where you would like to work is six times as effective as mailing out résumés and cover letters.

Richard Bolles
What Color is Your Parachute?

For those people who have gone through the career planning process step-by-step, the stage of focusing on specific opportunities is both exciting and fulfilling. After undertaking some sincere self assessment and exploring the various fields and areas of work, they have reached the stage where they can target specific companies and openings with confidence and optimism. They see the light at the end of the tunnel.

For those people who have skipped directly to the process of focusing on specific companies and openings, the light at the end of the tunnel looks more like an oncoming train. Because they have not carried out much research about themselves or their opportunities they feel great uncertainty about which direction to take. Often they respond by applying to everything in sight, creating a blizzard of laser-printed resumes and cover letters. This works about as well as standing in the middle of the San Diego freeway with a sign saying, "Hire me, I'm smart."

While targeting your job search to specific fields and companies does reduce the number of resumes and cover letters you might send out, it vastly improves your chances. Thus, the job search can occupy a smaller part of your daily routine (and a lot less of your present employer's paper supply). For the best prepared, who have extensive networks of people and sources of job information, attractive opportunities are much easier to find and may even find you.

How you learn about specific openings is very much dependent on the career field you are targeting. In academia and research, job openings are posted in a few, well-known publications and on the Internet. Practically everyone hears about them at the same time and practically everyone has the opportunity to apply. This is not the case in the business world. For example, large companies may have opening hotlines and internal postings, but many do not bother with advertising in the newspaper or other publications. To find out if Exxon or Sun Microsystems has openings you have to know the right number to call. Small companies may not need a human resources department and rely on hiring through their own network of people and the Internet. Federal and state governments post openings in very specific places and often you must submit an extensive government form in addition to your job materials.

Seeking Your Dream Job

Imagine your DREAM JOB. It's the place where the people are great, the compensation is excellent, and the challenges and the freedom know no bounds. You wake up every morning eager to get there and you leave at the end of each day satisfied and stimulated. Wouldn't it be great to have a job like that?

I know what you're going to say: I am a scientist! Any employment at all is a blessing!

Indeed, the abundance of Ph.D.s searching for gainful employment in science has led many scientist job seekers to ignore loftier goals for employment and stick to a strategy of simple survival. Get a postdoc, ANY postdoc; get a job, ANY job.

Before you succumb to such desperate thinking, realize this: there IS that PERFECT JOB waiting somewhere for you. You may not find it on the first fishing expedition, but that does not mean that you never will. However, your chances of finding the PERFECT JOB will be next to zero if you don't adopt a long-term strategy in your career development and your job search. There are ways to zero in on that perfect job if you prepare yourself well and know how to look.

Remember Your Passion

Perhaps the most important part of looking for the perfect job starts with an understanding of who YOU are and what YOU most enjoy doing. In Chapter 5 we discussed the various techniques of self-assessment and how to use those results to guide your exploration. However, many people can forget about those issues in the throes of the job search. Don't do it. Not everyone with a Ph.D. or a Master's degree in science has the same set of values. Some people love doing science and are happy to forego a high salary and a fancy office in exchange for intellectual freedom. Others hate working alone at the bench and would prefer team-oriented projects. Others have serious family considerations that they are not willing to push aside in order to get tenure. Different people are after different things in their careers.

Remember: if you don't like what you're doing for a living, you probably won't be very good at it.

Seeking an Employer of Choice or Fun Place to Work

The other side of professional fulfillment is finding a great place to work. It does you no good to find an ideal job if your coworkers are morons and your boss is a jerk! Most job seekers (scientists included) think that such factors are unknowable before you accept a job. This is simply not so. There are all sorts of ways that you can find out if the work environment is a good match for you. Be on the lookout for an employer of choice (EOC) or a fun place to work (FPW).

An EOC is an organization that is widely recognized to be a good employer and one that top applicants, people who have a choice, end up choosing. An EOC is a great place to work because your colleagues are likely to be of the highest caliber. These are the people from whom you will learn the most. In addition, having an EOC on your resume means that you will be more marketable in the future. How do you find an EOC? Just ask around. What institutions command the best reputations? Where have top people gone in the past? One thing scientists seem to have no problem with is rendering opinions about which institutions and companies are better than others!

An FPW is, as it's name suggests, an organization whose environment and lifestyle is fun for its employees. The work is engaging, the people are friendly and enjoy what they are doing, and the management knows how to create a productive and enjoyable atmosphere. Often an FPW is involved in interesting work, is a young organization, and has interested and committed employees. Every employer will advertise themselves as an FPW. Finding the ones who are telling the truth requires some research.

Sleuthing an Organization

Finding out what a company is really like requires some research. It is important to know some facts before you start asking questions. The company Web site is extremely helpful. In addition to finding out what the company does, how big it is, and what sort of growth it is expecting, some companies post the text of speeches and press releases from the senior leadership along with annual reports. These documents can provide great information about the atmosphere, areas of growth, and future direction of the company.

Informational interviews (see Chapter 6) are a great way of getting inside information about an organization, seeing the work environment firsthand, and getting to meet some people. You can tell a great deal about a company in a single visit just by talking to a few people and observing how they are working. Is the environment pleasant? Are people running around harried and freaked out? Are people too busy to talk? These are all indicators of the environment.

The other source of "insider information" that may be even more candid than an informational interview is someone you know who works in the organization. In most cases, you will not know anyone in an organization directly. But the people in your network probably do. These personal channels may end up giving you an earful about what the company is REALLY like on the inside. Anecdotes, from the people you speak with and from the corporate literature, provide a valuable window on the organization. Which people are identified as "heroes" of the organization and what did they do to become recognized as such? What are the common elements for success in the company? Threading all these bits of information together can give you a rich view of the company as a place to work.

Finally, don't neglect easy opportunities to learn more about your target organizations when they come to your campus or institution. Companies often visit to showcase new products, and this is an excellent time to button-hole someone and ask them about their workplace. In many universities recruiters may come to give information sessions to select groups of students such as MBAs. Don't be afraid to weasel your way into such events, even if you don't have an MBA!

The Final Element: The Potential Job and the Potential Boss

Even in an EOC or FPW, some jobs are better than others. The specific job is, of course, a major consideration. However, an equally important consideration is the boss, PI, or team leader.

A successful manager in a high-tech firm once gave me surprising advice about looking for the right job: "If you're applying for a job for which you are fully qualified, you are applying for the wrong job!" What she meant was—the best job opportunities are those that require you to stretch and learn new things. Furthermore, the best bosses tend to be those who hire smart, motivated people, rather than people who simply have all the requisite skills.

It is often hard to tell from a job description which assignments will give you opportunities to learn a lot. A job with fixed assignments that involves little interaction with others may not give you nearly as many opportunities to learn as a new position in a growing organization. Even in an established organization you should look for where the growth is: you will often find new job opportunities that will be defined by the job holder as much as by the management. It is possible to turn an OK position into a great job if the organization gives you some flexibility in defining the job for yourself.

The other important consideration is your boss. Perhaps the rarest element in the entire universe is the GREAT BOSS. A person who is a natural leader, enjoys mentoring, and is committed to success is a wonderful person to work for. Unfortunately, not only are such people rare, they tend to advance rapidly. Nevertheless, these people are out there—and you only have to talk with a few of their employees to find out who they are.

Ingredients for the JOB FROM HELL

Just as there are perfect jobs out there for you, there are also positions that would be exactly the WRONG thing for you. While this depends a great deal on who you are, it also depends on the job, the boss, and the organization. Here is my list of five ingredients for a perfectly hellish job:

1. *Bad management*—People who have no vision, poor communication and organization skills, and lack enthusiasm for the mission of the organization are surprisingly abundant. Figure out who they are ... and avoid them!

2. *Bureaucracy*—Be on the lookout for a process-laden organization. Often the absence of a profit motive can allow companies to bloat to the point where it is difficult to get even the most mundane task accomplished.

3. *Lackluster colleagues*—Just as in graduate school, you will learn more from your coworkers than you will from anyone else. If you visit an organization and come away feeling that you would be the smartest person there, beware!

4. *Poor resources*—Even the most dedicated and motivated workforce will eventually fail if they lack the resources to get the job done. Slim budgets, unstable funding, and second-rate equipment all spell trouble.

5. *Stress*—The above four factors can all conspire to make an extremely stressful work environment. A little stress is not a bad thing, especially if it is for a good cause. Constant stress isn't good for anybody, and it's a symptom that the underlying health of the organization may be at risk.

How Do You Find Out About Specific Openings?

Your Network

If your network is well-established and large, it is likely that you will learn about many jobs. In most cases, this will be insider information, and you have a distinct advantage over other candidates when and if the position is ever publicly advertised. You will also have an advantage because your name will have come "through the grapevine," and implicit in that is the recommendation of the folks who gave you the information in the first place: "Oh, she's a friend of Sandy's."

Through the Traditional Channels

Depending on the industry, you will be checking the traditional sources of job information regularly, either by reading a trade publication; reading, calling, or logging on to a job listing service; or through recruiting programs. When you are unemployed or soon-to-be unemployed, these activities are a high priority in your work day. If, however, you are presently employed, you may not follow the listings with much vigor. There is some danger in deferring this activity until you really need it. For starters, by looking at job listings even when you are not actively looking for a job you learn who is hiring and who is not. Maybe you're interested in Schlumberger, a leading geophysics corporation? A lack of posted openings for 3 months may be telling you something—either that they are not hiring or that you are looking in the wrong place. Making an effort to read job listings in your areas of interest, even when you are not actively looking, is a way of gauging the entire industry, not just the job market.

On the Internet

As we discussed in Chapter 6, the Internet is an extremely powerful tool for gathering information about potential employers and opportunities. Not only can you get job listings, you can get the latest information about the company's performance, current projects, goals, and corporate philosophy. Knowing all this before you apply for an opening will give you an edge.

Specific Job Resources on the Internet

Company listings

Many companies post some or all of their current openings on the net and these lists are updated at least as often as printed or telephone listings.

Random postings

Occasionally a company will post an advertisement on a newsgroup, chat group, or electronic bulletin board that they think has people who might fit the bill. If you see one, consider that the company that sent it already thinks people like you might be a good match for the position.

Job listing services

In some industries, such as programming, information technology, and—believe it or not—the hospitality industry, there are general lists of postings. Some of these are maintained by individuals, others by companies.

A Cautionary Note About Resume Banks

Resume posting services are a mixed bag. In a few industries (programming and information technology) the resume listing services are well established and used by employers. However, in most other cases they are a waste of time. For starters, realize that resume posting services bypass the sort of career planning and focused searching that we've been talking about in this book! Rather than being proactive and seeking out employers, you are passively posting your materials, allowing others to choose you. This does not demonstrate initiative or directedness and can leave you "kissing a lot of frogs" before you ever find a prince.

When employers use these resume databases, they tend to be filling lower-level positions, not positions that call for the unique capabilities that you possess. And employers tend to be trawling only for people who have "hot" skills, such as programmers. Most scientists have a richer set of skills, and you should be careful how to craft your resume to show this.

There can also be a more insidious use of resume databases: people set up fake resume databases, get the personal (and in some cases, confidential) information from thousands of resume posters, and then sell the information to direct marketers. So unless you want to start getting a lot of solicitations for subscriptions to *Popular Mechanics*, I would research the resume posting service very carefully before submitting materials.

Get Inside Information: Ask Questions

It is surprising how passive most people are during the process of applying for job openings. Most people prepare their materials with only the information from the job listing to guide them. This is unwise. Job listings are never the whole story. For starters, employers can't make ads too long because space costs money! What is more important, in any posting, what is described in the advertisement may not really describe the underlying needs and concerns that led to the opening in the first place.

Learning more about the circumstances, issues, and caveats that underlie a posting is enormously beneficial for an applicant. For starters, additional information allows you to craft a cover letter and resume (or CV) that is a better match to the needs and concerns of the employer. Second, contacting the employer demonstrates your interest and the ability to do your homework. Sometimes a simple phone call requesting more information ends up being a telephone interview.

Consider the situation from the employer's point of view. Filling an opening is probably one of many tasks they are trying to accomplish, and they want to do it as quickly and efficiently as possible. They are looking for a good match. By providing you with more information they are constructing a better applicant. They need you as much as you need them.

This advice is particularly important for positions in research or academia. Often, the posted listing represents some compromise of interests among the faculty or research staff. By learning more about the underlying needs, concerns, and goals of the individuals on the search committee, you will be able to present your best side. For example, a friend of mine saw an ad for an experimentalist in an Earth Sciences department at a major research university. Taken at face value, he would have applied to this opening in the same old way. Instead, he called a friend in the department who did some sniffing around and reported back to him that there was an insider from another major university that they were trying to attract. This department had about $1 million of start-up money and some lab space that they had to use that year. My friend refashioned his CV, cover letter, and research statement to demonstrate that he could construct a quality research program with those resources. He aimed to place himself as the first runner-up, knowing that fewer than 50% of the attempts to lure faculty away from other schools succeed.

Getting Experience While Looking for a New Career

Up to now we have been talking about information gathering. Another means of exploring new careers is by **doing**. More and more people are trying out different careers by volunteering, moonlighting, part-timing, interning, or simply insinuating themselves in other jobs. People gain experience and exposure by actually trying out different careers. And, depending on the arrangement, they may earn some spending money as well. Usually, people use this strategy when they have identified a particular field or company that they are interested in and want to improve their chances of being hired for the next opening.

Taking the concept of volunteering too far.

Volunteering

Volunteering can be the easiest and least time-intensive way to learn about an organization. Volunteering is easy to do for non-profit and community-benefit organizations because they have firmly established programs for volunteers and often rely on them for important parts of their operations. Offering your time to a private sector company is more uncommon and is usually difficult to arrange, but in some cases it can be done. Sometimes these organizations refer to the volunteers as interns. It is a good idea to ask about interning in an informational interview. Volunteering is also advantageous for those who are already working because the organizations to which you donate your time are usually flexible about when and how much you work.

If you are in a university, there may be some excellent volunteer opportunities right on campus. For example, if you are interested in technology business development you might consider volunteering at the campus technology transfer office. Simply organizing your department's speaker series can give you great networking opportunities, provided you run the seminar well!

Part-Timing, Moonlighting, and Consulting

Part-time jobs are excellent ways to cultivate new skills, increase your income, and work your way into a new career. However, it must be done with discretion. Many graduate students are on fellowships that do not permit them to work outside of school. Even for those who are only supported part-time there is a general expectation that the other 150% of your time will be spent doing your research! Advisors are often very reticent to give their approval for outside work—unless a graduate student can support himself or herself no other way. To be fair, advisors want to keep their students on track and prevent detours that will delay completion of the thesis. This is a serious consideration, and you should discuss the issue with your advisor or department chair before starting any part-time work.

So what do you do if you really want to work outside of graduate school and your advisor won't permit it? One option is to do it anyway. This is called moonlighting. I bring up the subject of moonlighting because it is surprisingly common in graduate school and is almost never discussed.

Moonlighters run the risk of being caught and, in an extreme case, being kicked out of school. However, some students find that they simply have no other way of gaining the skills and connections they need to advance their careers. I believe that students have to accept full responsibility for their career development. Explicitly countermanding an order from your advisor NOT to engage in outside work is simply unethical, but then, so is using graduate students as slave labor. My advice here is to proceed with extreme caution—sometimes it's better to ask forgiveness than permission.

If you are thinking of trying to set up some outside work, think of how you can frame the project or projects in terms of "consulting." Consulting is the process by which you hire yourself out as an expert for a specific project. Consulting is an activity with which faculty and professional scientists are familiar. Thus, there are some fairly well-established rules about not using work resources such as computers or lab equipment for outside work. As long as you follow those rules you should have no problem. If your advisor or supervisor takes issue with your consulting you should be respectful but firm: know your rights and responsibilities. In many cases, having outside consulting experience is a plus on a CV as well as on a resume. Your superiors do have a right to your undivided attention for the portion of the day that they pay you, but they do not have the right to restrict your professional development by threatening to fire you for doing outside consulting. Rules within individual organizations may vary, but very few graduate students or postdocs sign a contract at the time of their arrival that restricts their rights to consult on their own time using their own resources.

The Internet has ushered in a variety of new ways of freelancing and working part-time. Web sites such as FreeAgent.com, eLance.com and FlexMind.com all make a business of matching freelancers with temporary jobs that they can do remotely. Some companies are assembling virtual teams to complete entire projects, then releasing the team members when the job is finished.

Internships

Internships are the most time-intensive means of trying out a different career. At the same time they tend to be the most value because:

- You work full-time.
- They are structured for beginners.
- They usually provide ample opportunities for mentoring.
- Many are intended to cultivate future employees.

Internships and mentorships are most common in research settings and government agencies, but many exist in the private sector. Most are geared toward undergraduates in their junior or senior years, but there is usually no real barrier to graduate students applying for a summer. Recently, a few summer internships aimed at Masters and Ph.D. science and engineering students have appeared. For example, First Boston Securities advertised for science and engineering Masters and Ph.D. stu-

dents who wanted to try their hand at investment banking for a summer.

Science policy is one area that has several excellent internship and fellowship opportunities. Scientists are most familiar with the AAAS Congressional Science Fellowships, which places young scientists as staffers in Congress or in a number of federal agencies. There is also the White House Fellowship Program, which places early to mid-career professionals in all fields—including scientists—in high-level positions in the Executive Branch for one year. Much larger than either of these is the Presidential Management Internship (PMI), a program that recruits over 200 promising recent graduates each year for positions in government. Unlike the fellowships mentioned above, interns in the PMI program are given permanent positions in their host agency at the end of their 2-year internship. Finally there is the Christine Mirzayan Internship Program at the National Academy of Sciences. This internship engages graduate science, engineering, medical, veterinary, business and law students in the anaylsis and creation of science and technology policy.

Incorporating the Outside World in Your Research

Maybe there is a way that you can both gain experience in an outside field *and* maintain your commitments to your current employer. Some graduate students are gaining outside experience by incorporating issues and techniques from other industries into their research. For example, a geochemist I know was so interested in environmental policy that he structured his Masters thesis to approach both the science *and* the public policy of mercury contamination. Another, an astrophysicist, purposely incorporated neural net theory and computational techniques into his postdoctoral project, gaining him experience in a field of applied mathematics with many applications in financial modeling. In both cases, the efforts these individuals made led to job offers and exciting careers in which they are recognized as experts.

These are only two examples of the many possible bridges you can build between your research and the outside world. No only will it benefit your research and your career but the connections you make and the techniques you learn will probably bring long-term benefit to your advisor, department, and school. Already, there are a few graduate programs that are trying to formalize the connection between basic research and practical applications. By taking the initiative, you will set yourself apart in both the world of research and the outside world of work.

Summary

- Focusing on specific opportunities should come after you have completed some self-assessment and have explored the broader job market.
- There are great jobs and horrible jobs—you can find out which is which by remaining observant and talking to people who are familiar with the company or organization.
- You can become an insider by volunteering, interning, consulting, or moonlighting with a prospective employer.

There IS a difference.

CVs and Resumes (There IS a Difference)

9

The best resumes say as much about where you are going as where you have been.

Al Levin, co-author
Exploring Your Career Beliefs

Typically, people begin the process of searching for a job by constructing a resume or curriculum vitae (CV). Most of you probably have at least one or the other. Some of you may have even consulted a book or two about writing resumes. However, most of you may have received no formal advice and are proceeding based on the examples of friends or family.

The rules of CV construction may be familiar to some of you, since we come from a work environment where that document is the norm. Most research scientists, however, do not have any experience writing a resume and wind up creating a document that looks very much like, well, a CV. There are some fundamental differences between CVs and resumes—both in style and in purpose. However, despite these differences, the goal is the same:

The main purpose of a CV or a resume is ***to get you an interview.***

This may seem like only a distinction in semantics. However, it's a critical concept, one that trips up many job applicants. A resume or CV is used by employers principally to develop a *short list*. Once that short-list is complete and candidates are brought in for interviews they are judged on a substantially different set of criteria (more on this in Chapter 12). A common mistake that many job seekers make is to focus mainly on their resume and neglect other important aspects of the job search, such as networking and researching opportunities. It can be very satisfying spending a weekend tweeking the fonts and margins of your resume but it is likely that the time would have been much better invested by going on an informational interview, browsing the Web for opportunities, or gathering information from your network about a particular opening.

OK, so what IS the difference between a CV and a resume?

The CV is a summary of all your educational and professional background. It is used when applying for academic jobs (in teaching or administration), for research jobs in government or private laboratories, or for a fellowship or grant. A CV commonly includes a full list of publications and can be several pages in length. Space is not at a premium. The CV will be thoroughly examined by at least one member of the search committee who will glean the information that is of highest importance to the committee.

The resume is a summary of those aspects of your job experience and education *that qualify you for the particular job for which you are applying*. Read that again. The big difference between a resume and a CV is that you tailor the resume to the specific needs of the employer. You may alter the order in which you present skills or experience, you might toss out irrelevant skills and experience altogether. The critical issue is that you need to have a clear understanding of the needs of the employer or the industry before you can do this. This requires research.

Resumes are used everywhere a CV is not specifically requested. A resume is shorter, usually 1 page, sometimes 2 for higher-level positions. Space is at a premium, and the layout is compact but easily readable. The *average* employer spends about 20 seconds examining and sorting each resume. Thus the challenge is to provide the right information in an attractive layout that will cause the employer to place your resume in the "for closer examination" pile rather than the "maybe later" pile.

Resumes from research-trained scientists applying to positions that do not call for an advanced degree are often in danger of being immediately discarded. Why? Because most employers tend to believe that Ph.D.s are overqualified, inexperienced in the real world, and best suited to, well, research! The biggest challenge for scientists applying for non-science jobs is convincing employers in the first 20 seconds that you are serious, competent, and prepared, and not over-qualified, misdirected, or out of place. Your resume, like your cover letter, your interview, and every other part of your package of job materials, must reinforce the preconception that you are bloody brilliant (after all you are: you are trained in science!) but must also immediately challenge stereotypical notions that you are a loner, impractical, and uninterested in things besides science.

You can also understand how important it is to try to make some inroads with the employer before submitting your resume. One great way is to have your resume referred to the employer by someone else in their organization. Who might do this for you? A member of your network or someone with whom you have had an informational interview in the past. If you have no personal connections with the prospective employer, an initial inquiry, good cover letter, and follow-up e-mail can help your unusual resume be interpreted in the right light.

If you have read everything up to now you may realize that it is critical to undertake some sincere self-assessment before trying to construct materials that you will send out to strangers. Self-assessment is particularly important if you are exploring alternatives to research science. After all, a resume should say at least as much about *where you are going* as it does where you have been. You may also realize how important it is to have carried out exhaustive research on the organization and opportunity to which you are applying.

Your resume is your only bullet: Don't fire it until you have aimed as precisely as you can!

In the past, people have discussed converting a CV to a resume as if all that mattered was having a single page and the right font. No way. Depending on how far you are moving away from a career in research science your resume will bear less and less resemblance to your CV. Your technical and research experience will be a major item or series of items in your resume, but you will likely include other things as well, such as relevant skills and experience from the rest of your life. And, yes, even for the most die-hard among you, you *do* have other parts of your life besides science. Later in this chapter and in Chapter 10, we'll discuss the value of highlighting non-technical accomplishments in your resume.

Basic Parts of a CV

The overall layout of your CV is NOT going to make much of a difference in whether or not you make the short list. However, presenting a CV that is organized, attractive, easy to read, and contains all the information of interest to the search committee will help them evaluate you. Making their lives easier can't hurt, now, can it? CVs that are missing critical information, are poorly laid out, or are hard to read can only hurt your chances.

The contents of your CV may vary somewhat, depending on your field. If you have not already done so, check out your advisor's CV or that of a collaborator to see what categories of information are included. Some categories of information are essential in all fields. These are:

Identifying Information

Usually your name, address, phone/fax number, and e-mail address come first. Be sure to include your citizenship AND make sure your name appears on each page (either in the header or footer of the CV). DO NOT include your date of birth, marital status, number of children, social security number, or other personal information. Not only is it not required, but it is illegal for employers to request this information of you (more on that in the resume section).

Education

List your degrees, along with the department, institution, and dates of completion or expected date. In each, list minors, sub-fields and any academic honors such as Summa Cum Laude. Some people list the titles of their theses and the names of their advisors as well; others have a separate section for this.

Dissertation or Thesis

Some people have a separate section giving the title and a brief, one-paragraph description of their work. In some disciplines of science, it is customary to simply list the title and describe the research more fully in the experience section.

Awards, Honors, Fellowships, Scholarships

All awards outside of academic honors should be listed in this section. Honors such as membership in Phi Beta Kappa, outstanding paper awards, and competitive fellowships and scholarships should be listed, along with the year.

Professional Experience

This section should be used to describe all your past positions and experience. Some people structure this chronologically, giving the position, such as Research Assistant, along with the institution, the date (usually month/year), and a brief description of the activities involved. It is important to list more than only job titles. Explain what you did in each position. Don't assume that those reading your CV will know exactly what went into your year as a TA. Be specific, be brief, and *quantify your accomplishments as much as possible.* Give numbers, cite examples, and be specific. Some people break this section up into subheadings such as Research Experience, Teaching Experience, Consulting, and others. If you do this, make sure that the information is still easy to find.

Publications

This section is usually at the end of every CV, usually because it alone takes up a few pages. While citation styles may vary from field to field you should keep one thing foremost in mind: make your publication list as clear and easy to understand as possible.

Organizing your publication list into sub-categories of importance is one way of clarifying your publication record. Peer-reviewed full articles are by far the most important part of your publication list and should appear first. Publications in press or in review should be listed separately. Conference proceedings, reports, and other publications that are not peer-reviewed should be next. Other publications, such as published photographs or patents, can be listed in their own categories. Abstracts are the least important and should be in their own section at the end. Some people list only the last few years' abstracts to save space.

For scientists with several pages of peer-reviewed papers in print, the list of full publications itself can be sorted. For example, those of a crystallographer might be sorted into fields of Solid-State Physics, Materials Science, Ceramics, and General/Interdisciplinary. Any format that helps people assess the quality and breadth of your record is an asset.

In contrast, layouts and citation styles that tend to obscure the quality and breadth of your publication record should be avoided. Mixing citations for abstracts and full papers together only forces people to hunt through the list and find the few items that are full articles. Citations that obscure the order of authorship or leave out some information only waste time. In some cases, these styles can leave the impression that the applicant is trying to camouflage an inferior publication record. Despite what you may think, duplicity is NOT a valued trait in academia.

Other Sections

Academic/Professional Service

Some people have worked on committees, volunteered in educational projects, or served their school or organization in other ways. This is useful information, especially when applying for jobs that will value this type of service.

Memberships or Professional Affiliations

Many people list all the professional and scientific societies of which they are active members, along with the date at which they joined.

Grants and Funding

For some jobs, getting grants is not just laudable; it's essential. People with experience raising their own funding sometimes list their past and current grants in a section of their CV. Usually the title, agency, amount, and dates of the funding are listed.

Courses Taught

Some people list the course titles that they have taught. Other people list them in their Teaching Statement.

Students Advised

Some people who have experience advising students (usually in research projects) list the names of their past advisees.

Languages

Some people list the languages in which they are fluent.

Letters of Reference

Imagine a nightmare scenario. You have worked hard for years on a difficult project with a difficult advisor. At last the end is in sight—you're ready to defend your thesis and apply for jobs. But after months of time and tens of applications, you are still getting only rejection letters. You can't figure it out. Another member of your research committee finally bugs one of his colleagues until she gives you an interview. During that interview your prospective postdoc employer explains her dilemma: your record is good, you are a promising candidate ... but the letter of reference from your advisor is NOT.

This story is the reality for more people than you'd imagine. In some cases, the advisor and the student have had a truly terrible relationship, but in other cases the advisor has simply failed to provide the kind of endorsement that is suitable—or sufficient—for the job being sought. The result is the same: your job prospects are dimmed considerably.

Most of you probably think that the letter your advisor or boss writes is completely out of your control. After all, it is your advisor's opinion—not yours—and the letter is confidential. Surprise! You can actually exert a significant amount of control in the content and tone of the letters that are sent out on your behalf. With careful planning and good communication you may even mitigate the negative things that one recommendor may put in the letter.

What is a good letter of recommendation?

Having reviewed applications for a number of fellowships and several postdoc positions, I have seen the full range of letters of reference—from the stellar to the abysmal. You might think that the difference lies only in the range of praise given. It goes without saying that if your recommendors think you walk on water, your letters of reference will be pretty darn great. However, even stellar achievements might not be brought out as much as they deserve if the structure and tone of the letter is lacking.

The best letters of reference tend to have the following characteristics:

1. The writer is known and respected by the reader (either a colleague or someone well-respected throughout the community).
2. The writer gives specific and meaningful examples of achievements, and tells specific stories illustrating strengths of the candidate.
3. The writer provides quantified assessments of the candidate's abilities, especially with respect to other graduates.
4. The letter addresses the most important skills and traits needed for success in the job.
5. The letter is long.

For example, a strong letter might say things like:

> *Carol is an outstanding researcher, in the top 3-5% of the graduates from our institution over the last 10 years. She has the drive, creativity, and ability to become a leading research astronomer, even in this tough job market.*

In contrast, a weaker letter might read:

> *Carol has shown dedication and drive throughout her years as a graduate student. She has the capacity for continued productivity in the field of astronomy.*

I think it's fairly obvious which is the stronger endorsement.

Think Like Your Advisor

Consider the letter-writing process from your advisor's perspective. Most likely he or she genuinely cares about your success and is willing to write as positive a letter as possible. However, they are also busy and writing your letter of reference may be one action item competing with many others on their to-do lists for the week. Also, they may not have much direct information about the opportunities you are seeking, especially if they are outside of academia. They probably have a good idea how to write a letter for a postdoc position, but they may be clueless about what to say if their students are applying for a AAAS Congressional Science Fellowship or an internship. All this can conspire to make for a lackluster letter of reference.

The Keys to Superior Letters of Reference

Now that you've considered the letter-writing process from your advisor's perspective, the ways in which you can help him or her create a better letter should be obvious:

1. Give your recommendors plenty of time to accomplish their tasks. Believe me, telling them that you need a letter by the next business day will not only throw their worlds into turmoil, but it will also speak volumes about your ability to plan and organize! Do them and yourself a favor and give them plenty of advanced warning.

2. Remind them of the deadlines and check in periodically. Even the most well-intentioned recommendor may slip up, get confused on deadlines, or lose some information. Periodic reminders will NOT be resented—they may actually be grateful!

3. Prep your recommendors THOROUGHLY. One good strategy is to write them a memo describing the job or jobs to which you are applying. Include a copy of the job description and any other relevant material. In your memo tell them what key skills or attributes are important and what specific things in your background they should mention. Come as close to drafting the letter for them as you are comfortable with.

Know Their Mind BEFORE Asking for a Letter!

Ideally, it would be great if you could know what sort of letter your recommendor was going to write BEFORE you ask them for a letter. While you're not a mind reader, you probably have a good feeling for the health of your relationship with that person. This is the best indication of what sort of endorsement he or she will provide in the letter. However, there are strategies for assessing the quality of the letters they may write.

One strategy is the direct approach: sit down with your advisor, or whoever will be writing your letter of reference, and discuss your career plans. Ask your advisor directly if your career goals are realistic or in what ways you need to become more competitive. This may be a very difficult discussion to have but, believe me, it is much better to get these issues aired at the outset. At worst, you both may realize that there is a disconnect between your career goals and your advisor's preferences or your perception of your performance and theirs. At best, you may discover that your advisor is a lot more supportive of you than you realize, and he or she may have some great advice for you.

The other strategy is the indirect approach: talk to the most trusted mentor you have (maybe a member of your research committee or a collaborator from another institution) and ask him or her to "sound out" your advisor for you. In a single conversation he or she will probably get candid comments from which an assessment can be made. While it is unlikely that the intimate details of their conversation will be shared with you, he or she should be able to indicate how strong an advocate your advisor will be.

What if my advisor will NOT write a strong letter?

Many students worry that the disapproval of their advisor will sink their career on the spot. This difficulty can be mitigated with some careful strategy. First, it is important to understand the nature of your advisor's criticisms. You may get this either from a direct conversation or through indirect methods (for example, depending on the policies in your department, you may be able to see your personnel file with your yearly evaluations). Once you have a good handle on the issues, talk to another recommendor—someone you trust—and ask him or her to address these points in a letter. A second person can render a "second opinion" and in some cases make a stronger case for you by pointing out the sources of the problem between you and your advisor and how you handled it. As a person who has read several hundred letters of reference, I can say that this strategy can go a long way in mitigating the negative comments of a single recommendor.

Another strategy is simply to rely on letters from other people, and not your advisor. While this will raise questions in the mind of a potential employer, you should ask your other recommendors to explain the situation for you in their letters. More often than not, a potential employer will be reassured if the letters you do provide are uniformly positive and explain the nature of the conflict between you and your advisor. If you do this right YOU may come across as a far more mature person than your peevish advisor!

Teaching and Research Statements

There are many parts of the academic job search with which we are familiar. Nearly all of us have, at this point, a CV and a publications list. During our years in grad school, most if not all of us have listened to job talks from prospective faculty. However, one part of the academic job application process that does not seem to be discussed very often is the "statement of research or teaching interests."

Seeking the truth behind this unnerving yet time-consuming part of the academic job search, I consulted with seven present or former faculty members from a variety of colleges and universities in the United States. Combined, they sat on over 40 hiring committees.

Their advice about form and content (see below) is fairly consistent and probably won't be too much of a surprise to you. But what surprised me was their view of the relative importance of these statements in the overall hiring process: Teaching and research statements do not seem to be very important in most cases! None of my experts could cite a single example of a research or teaching statement being a pivotal factor in selecting a short list.

The factors that loom larger in the minds of hiring committees, chairs, and deans are more global. Areas of specialty, grantspersonship, overall academic prowess, compatibility with the department or school, and the ability to teach in required areas are of much greater interest. The statements of research and teaching interest are used to help assess these factors, but other elements of the academic job application are far more important, especially the content of the CV, the letters of recommendation, and the cover letter.

Even though these statements are less important than other parts of your application package, they cannot be ignored. Several professors noted that while teaching and research statements rarely help very much, they have the potential to seriously compromise a candidate. In particular, statements that show a lack of understanding of the type of school, the needs of the department, and the specifics of the job opening are more likely to hurt one's chances. With teaching and research statements it seems hard to hit a home run, but it is easy to strike out.

First, Some Definitions

According to the professors I talked to, as well as several publications on the academic job search (see below), a research or teaching statement should be one and a half to two pages in length. Styles can vary, but most sources suggested a readable, organized format that combines paragraphs and bulleted highlights.

The Research Statement (or Statement of Research Interests)

Your Research Statement should be a summary of your research accomplishments, current work, future direction, and potential. These statements are usually one to two pages in length. Your statement can discuss specific issues such as funding history and potential, requirements for laboratory equipment and space, and potential research and industrial collaborations. It should be technical, but should remain intelligible to any member of the department. Because it has the potential to be read by people outside of your subdiscipline, the big picture is important to keep in mind. The strongest research statements present a readable, compelling, and realistic research agenda that fits well with the needs, facilities, and goals of the department. You can substantially weaken your statement by proposing an overly ambitious plan, lacking in clear direction or big picture focus, or through inadequate attention to the needs and facilities of the department or position.

The Statement of Teaching Interests

This statement provides a summary of your teaching experience and subject matter expertise. It may discuss the courses and areas you are qualified to teach. Some applicants discuss pedagogy (teaching techniques and methodologies), especially if they have experience teaching with particular tools, such as computers, or teaching particular classes of students, such as those with disabilities. The strongest teaching statements demonstrate a depth and quality of teaching experience, an interest in teaching commensurate with the goals of the department and school, and clear competency in the subject matter areas specified in the job description. Teaching statements can stray by failing to provide specifics on subject matter expertise, or failing to demonstrate a sincere commitment to teaching (especially for applicants with research-oriented training), or through excessive pedagogical philosophizing.

The Three Most Important Rules for Writing Teaching and Research Statements

Know the School

Richard Reis's book, *Tomorrow's Professor,* has an excellent discussion of the environment of higher education. For starters, not all schools are alike. The Carnegie Classification categorizes U.S. schools into 11 categories. Most salient to the aspiring science professor are:

Research I and II schools
Doctoral I and II schools
Masters I and II schools
Liberal Arts I and II schools

Each type of school differs greatly in its levels of external funding, balance of teaching and research, size, and number of degrees granted. However, most academic job applicants come from Research I and II institutions. Understanding and appreciating the different goals, missions, and environments of an institution is critical BEFORE putting together your statements. For example, a number of Masters I and II institutions are more focused on teaching and accrediting secondary school teachers than on the creation of basic knowledge. If you apply to a Masters I and II school for a job, your statements had better acknowledge and embrace these goals.

Know the Department

Job ads never tell the entire story. Learning more about the needs and priorities of the department, as well as the circumstances surrounding the opening, can give academic job seekers a big edge. This is where your network can be so valuable. If you do not have a friend on the faculty or in the student body, perhaps someone in your network does.

The more you learn about the department and the faculty, the better you can assess how you might fit their needs. By learning about and addressing their priorities, whether or not they are stated in the ad, you can greatly differentiate yourself from the rest of the pack. For example, is undergraduate teaching a growing priority? Is the department concerned about the number of undergraduate majors? Is the department looking to increase its prestige? If your research and teaching statements address these priority issues, your application will probably be read with great interest.

Know the Position

Some academic positions are created to fill specific needs. Some job openings are "superstar" hunts, designed to attract rising hot shots and raise the profile of the department. Often you can tell the difference simply by reading the advertisement carefully. Other times it may be less clear. Again, your network can help with this!

Finally, it is important to know when an opening is REAL and when it is not. Sometimes, departments will advertise an opening or an "anticipated" position in order to gather a huge stack of CVs to wave under the nose of their dean to convince them to actually authorize a position! These don't always pan out. Another occasional pitfall is the existence of an insider. The only consolation in these circumstances is that sometimes the insider stumbles or ends up taking another position. One of the faculty members I talked to estimated that this happens at least 30% of the time. If you are the #2 pick, you may still end up with an offer.

Teaching and Research Statements, the Final Frontier: Tenure

The teaching and research statements you write as a job applicant might not hold that much sway during the hiring process. But those you write as part of your tenure review package will be VERY important. One professor I spoke with told me she kept building and perfecting her statements of teaching and research interests every year that she was an assistant professor. By the time she was ready to put her materials together for her tenure review, she had a dynamite document. And, yes, she WAS granted tenure!

Cover Letter

As with resumes, cover letters that accompany your CV should be tailored to the job to which you are applying. A cover letter should:

- state the specific position to which you are applying
- explain how you learned about the opening
- accent your most important qualifications

The cover letter should accent those skills and experiences that are of greatest relevance to the job. If you are applying to a small liberal arts college you may want to emphasize your experience teaching and advising undergraduates. If you are applying to a major research university you may want to point out that you currently have $330,000 in research funding from NASA. We will discuss cover letters more in Chapter 11 but an additional resource that discusses cover letters and gives some good examples is *Cracking the Academic Nut,* by Margaret Newhouse.

Resumes

There are two general types of resumes: chronological resumes and functional, or skills, resumes. Chronological resumes are what you are probably most familiar with; they list your work experience in chronological order. Functional resumes categorize your experience under several key skills areas: those needed for the job you have targeted. Chronological resumes are useful for demonstrating a pattern of working, especially if you are continuing in a general profession or field. They emphasize progression and a steady history of work. Functional resumes are structured to emphasize marketable skills. They tend to be more effective for people switching to new career fields or for people who have worked off and on for some time (see the mostly true story of Karen Smote in Chapter 10 for an example of this) because they de-emphasize the mismatch of past work experience and gaps in work history.

Most people use a combination of these two styles in which work history and relevant skills are presented. Work experience is listed, usually by job title and in chronological order, but this is followed by a description that emphasizes the skills used.

Basic Parts of a Resume

Name and Address

This part is easy: put your name, address, phone number, fax number, and E-mail at the top. If you really want to show off you can put the URL for your home page, too! If your resume is two pages long, be sure your name is in the header of the second page.

Objective Statement/Summary Statement

One way to clarify your specific career goals and qualifications is to put an objective statement or summary statement at the top of your resume. These are usually only one or two sentences. Think of it as a "mission statement" or a succinct advertisement of yourself.

An objective statement shouldn't be too narrow, too broad, or too vague. Saying something like "applicant desires a challenging position utilizing his skills and experience with the opportunity for advancement" would tell an employer that you wanted a job, any job, and that you have no clue what you want. Not the best first impression to make.

Here are some clear, concise objective/summary statements:

> Seeking entry-level position as Web page designer or HTML programmer for technology start-up or Web design consultancy.
>
> Position as associate in strategic management consultancy specializing in biotechnology. Six years experience in biotechnology research and marketing.
>
> Position as analytical chemist in semi-conductor manufacturing company, specializing in transmission electron microscopy.

Each of these clearly states the employment goals and key experience of the applicant. You can clearly see that, in order to construct these objective statements, the applicants would need a very good understanding of the nature of the job opening.

When to use and not use a summary statement

Over the past few years employers have found less value in objective/summary statements. Most are hopelessly generic and do not add much value to the resume. In some cases the applicant is applying for a specific position and it is unnecessary.

Before you include an objective/summary statement on your resume, carefully consider its value. In some cases such a statement can be very helpful. For example, if you are sending out a resume to members of your network, including this statement can help them, and the people they pass it to, understand your exact goals. However, if you are applying for a specific job opening, this statement might be redundant.

Education

The educational background of research-trained scientists is usually outstanding on paper. It is something that people will really notice. A candidate who has a Ph.D. from MIT in Geophysics and graduated Summa Cum Laude from Vassar College will cause anyone to sit up and take notice. In fact, *any* advanced degree in a resume submitted for a position for which an advanced degree is not the norm generally should be thought of as an asset (but see the story of Karen Smote for a counter example). Put the Education section either right under the Objective/Summary statements or at the bottom of the page. Be sure to include academic honors like ___ Cum Laude in this section, but put other honors and awards in a separate section. Just so everyone is clear on this, you should put the following in the Education section in reverse chronological order:

- Name of institution (Ph.D., Masters, Undergrad)
- Location of institution
- Year of graduation (don't bother with the month)
- Department or major (or dual majors) and academic honors (Cum Laude, etc.)
- Any professional certificates, accreditations, or minors

Do NOT bother putting in:

- The titles of your theses (that might go in work experience, but only if it is specifically applicable to the job opening)
- The name of your advisor
- Your GPA (if it is requested, often along with GRE/SAT scores, list it/them separately)
- Your high school

Some Masters and Ph.D. scientists have reported that they were turned away from jobs because they were "over-qualified." Some have suggested that, in some cases, you should remove Ph.D. from your resume altogether and pretend that you never went to graduate school (would a stint in prison look better?). If you feel this way, I suggest you reread Chapter 2. A Ph.D. or Masters is a liability only *if you are unable to show a prospective employer the valuable transferable skills you have acquired along the way.* If an employer cannot recognize the value of an advanced degree, they lack any imagination or business sense and would probably be miserable to work for anyway.

Work Experience

This is the place to put down three to five experiences/jobs that highlight the set of skills that are *most desirable to the employer.* These accomplishments should sound substantial and important. They should highlight your skills and talents. Most important, they should bring out the benefit that you brought to each organization and *show how you made a difference.* With some categories this might be difficult (such as teaching experience) but in others, such as research experience, you can really make yourself shine. If you have not already done so, go back and reread Chapter 5 and do self-assessment exercise #3. Do it!

Describing these things should involve using action verbs (see the list that follows) and in an active past or present tense. For example, rather than saying "was responsible for operation, maintenance, student training, and certification of users for X-ray fluorescence spectrometer 1992-1995," say "maintained and operated X-ray fluorescence spectrometer, trained and certified 44 students over 3 years." By using action-rich verbs and numbers, you highlight your accomplishments in quantitative ways. Grammatically speaking, one should spell out numbers from one to nine, and write out numbers like 23. However numerals tend to stand out and, for that reason, should be used even for single numbers (i.e. "3" instead of "three"). This latter point of quantifying your accomplishments is very important. Again, it may seem impossible with some topics, such as teaching experience, but if you can at least mention enrollment numbers, it is an advantage.

Unpacking Your Thesis: Using Action Verbs to Describe Your Experience

Opposite is a list of descriptive action-rich verbs that can be used to describe various aspects of your technical training and experience.

Other lists of action-rich verbs can be found in *The Damn Good Resume Guide* by Yana Parker and *What Color is Your Parachute?* by Richard Nelson Bolles.

If you are just emerging from grad school, your school research experience may be the first and biggest item, but it shouldn't be the only one. Teaching experience can look good as a separate category, especially if you had real teaching duties as opposed to grading the problem sets from your advisor's class. Summer work for companies or part-time work done while in school is a real asset. If you did something particularly notable in college, that can go in, especially if your work experience is limited—for example, being the technical director of a theater on campus. With each of these things you should list the following first as a heading:

- Job title
- Name of the organization
- Location (city, state) of organization
- Time of employment (again, use only years—nobody cares about months)

This information should be all on one line, perhaps in bold (again, see the resume case studies in the next chapter for examples).

Other Sections

You may want to include a list of particular skills if you have not already mentioned them in your description above. Computer skills and foreign language skills might go in this separate section. Depending on the job, you might want to mention particular software that you are familiar with. Since most of the "real world" uses C or C++ you should mention if you have some experience in these languages. FORTRAN is not widely used in the programming world these days. You might want to include a section on awards if they are particularly prestigious and recognizable to your intended audience.

Research-related
analyzed
assembled
built
charted
classified
collected
computed
correlated
created
defined
designed
detected
determined
developed
devised
diagnosed
disproved
dissected
engineered
evaluated
examined
explored
fabricated
integrated
investigated
mapped
measured
monitored
operated
photographed
prepared
proved
recorded
repaired
sorted
tested

Computer-related
calculated
compiled
de-bugged
installed
maintained
modeled
modified
processed
programmed
systematized
upgraded
wrote (code)

Communication-related
advertised
arbitrated
co-authored
composed
displayed
edited
illustrated
informed
interviewed
marketed
mediated
negotiated
performed
persuaded
prepared
presented
promoted
publicized
reviewed
sold
solicited
spoke
submitted
translated
wrote

Teamwork-related
assisted
collaborated
delivered
facilitated
helped
participated
recruited
referred
served
shared
supported

Project management-related
budgeted
conducted
coordinated
delegated
directed
headed
implemented
managed
motivated
organized
oversaw
planned
proposed
ran
reconciled
scheduled
selected
supervised

Teaching-related
advised
coached
counseled
demonstrated
developed
dramatized
encouraged
evaluated
explained
graded
guided
informed
instructed
lectured
presented
stimulated
summarized
taught
trained
tutored

Consulting-related
advised
assessed
audited
estimated
inspected
judged
predicted
recommended
reviewed
tested
verified

Verbs that convey accomplishment
accelerated
achieved
attained
boosted
completed
conserved
consolidated
corrected
discovered
distributed
doubled
exceeded (goals)
expedited
founded
improved
increased
initiated
launched
modified
obtained
reduced (problems)
resolved
revitalized
spear-headed
stream-lined
succeeded
unified

What Not to Include

It used to be cool to add some personal information—hobbies and the like. After all, maybe the reader is an avid hiker like you: dude, you've got it made! Well, this is a new century and personal information is not only extraneous—including it on a resume can seem unprofessional. Skip the little section at the bottom of the resume that says you love to ski, hike, shoot large animals, and collect spores, molds, and fungus. Also verboten are the following:

- Date of birth
- Your marital status
- The number of children you have
- Salary requirements

Superman updates his resume.

By law, employers in the United States are not permitted to ask you your age, marital status, or the number of children you have. They can ask oblique questions, such as "do you have any special needs that would affect your performance in this job?" You may think you're doing them a favor by volunteering this information, either in your resume or during an interview, but in reality, it gives them the impression that you don't know the rules and lack experience in the "real world" workplace. In Chapter 12 we discuss some strategies for dealing with inappropriate questions during an interview.

Including inappropriate information on a resume is a common mistake for young scientists new to the United States. The rules and requirements for resumes in Europe, Asia, and elsewhere are somewhat different. For example, in Europe it is appropriate to include personal information such as marital status and the number of children. In some countries in Asia it is legal for employers to specify gender when advertising for a job. Be sure you understand the particular rules wherever you are. If you are uncertain what is fair game in another country, contact a career planning or placement office or one of *Science's Next Wave's* foreign editors (www.nextwave.org).

References

References, if requested, should be listed on a separate page with their full name, job title, place of employment, relationship to you, full address, phone number, fax number, and e-mail. The goal is to give the prospective employer all the help they need in contacting your references. Making things easy for them will reflect well on you. Also, don't bother putting the statement "references available on request" in your resume; people know that.

References are not accorded the same weight in the outside world as they are in the world of research science. Most employers assume that anyone you list would be able to sing your praises (although hiring managers tell me it is surprisingly common to contact a listed reference who has little to say about a candidate!). Employers tend to rely more on the written job materials and the interview for making hiring decisions and consult references (most often by phone) as a final check. However, references that are known to the prospective employer can be extremely powerful. These people may get called early, and if they are prepared to sing your praises, you have a terrific advantage.

Do remember to prepare your references for the possibility of inquiries!

Preparing the Ideal Scannable Resume

As if the job search hasn't already become a humiliating and dehumanizing experience, some larger companies and government agencies are now using computer programs to sort through large numbers of applicants to find desirable employees. These organizations are taking advantage of a procedure called electronic applicant tracking. Basically, companies take your resume, scan it, and send the image through an optical character recognition program that plucks out the needed information and puts it into a database. An employer can then select a series of criteria to find the ideal candidate for a job (for example: FIND: EDUCATION=Ivy League, SKILLS=Overseas Marketing, SIGN= Capricorn). For example, the Clinton administration used RESUMIX, one such system, to cull the more than 5,000 applications they received regarding jobs in the new administration after the 1992 election. If possible, ask the contact person regarding the position if a scannable resume is recommended.

Why is it important for you to know this? When you prepare a resume for a company using this procedure, the resume must be scannable. A scannable resume has standard fonts and crisp, dark type. Scannable resumes must have plenty of facts for the obtuse little machine to extract—the more skills and facts you provide, the better the chances for potential matches.

The system can extract skills from many styles of resumes (the grammatical and linguistic search routines are surprisingly sophisticated). The most difficult resume for the computer to read is one of poor copy quality that has an unusual format, such as a newspaper layout, variable spacing, large font sizes, graphics or lines, type that is too light, or paper that is too dark. I saw a spectacular example of this. A friend of mine from grad school had submitted her resume to a large national laboratory. She had used a small font and lots of lines in her resume. The scanned version was a disaster: nearly every word was misspelled!

As far as content is concerned, employers search the resume database in many ways, searching for your resume specifically or for applicants with specific experience. In the case of the latter, they will identify key words, usually nouns such as writer, Ph.D., UNIX, Spanish, San Diego, etc. Here are some tips from RESUMIX for producing the best possible scannable resume:

Maximizing Scannability

- Use white paper, 8.5 x 11", printed on one side.
- Use laser printer original, avoid dot-matrix, photocopies, and poor quality typing.
- Do not fold or staple.
- Use standard typefaces such as Helvetica, Futura, Optima, Univers, Times, Palatino, New Century Schoolbook, and Courier.
- Use font sizes of 10 to 14 pt. (If using Times, don't use anything smaller than 12 pt).
- Do not condense spacing between letters.
- Use boldface and/or all caps for section headings as long as letters don't touch.
- Avoid fancy styles such as italics, underline, shadows, and reverses.
- When faxing, fax in "fine" mode.

Tips for Maximizing Hits

The database searches through all the resumes and ranks them by the number of "hits" or matches there are with the selected search fields and criteria. By this logic, the more information you provide, the greater the likelihood that your resume will get a hit and ascend the list. Here are some suggestions for maximizing matches.

- Use enough words to define your skills, experience, education, professional affiliations.
- Describe experience with concrete words (for example: "managed an analytical laboratory" rather than "responsible for managing ...").
- Use more than one page if necessary (the computer doesn't care).
- Use jargon and acronyms specific to your field (spell out acronyms for human readers).
- Increase your list of key words by including specifics, for example, the names of software you use, such as Microsoft Excel, Adobe Photoshop, etc.
- Use common headings such as Objective, Experience, Employment, Work History, Positions Held, Appointments, Skills, Summary, Summary of Qualifications, Accomplishments, Strengths, Education, Affiliations, Professional Affiliations, Publications, Papers, Licenses, Certifications, Examinations, Honors, Personal, Additional, Miscellaneous, References, etc.
- If you have extra space, describe your interpersonal traits and attitude. Key words could include "skill in time management," "dependable," "high energy," "leadership," "sense of responsibility," "good memory." See the list of transferable skills and traits to get some suggestions.

Should I prepare a separate "scannable" resume?

If you have submitted a resume for a specific opening, your resume will almost always be forwarded to the person making the hiring decision after being scanned. Thus, the resume you submit should be intended for human eyes. If you are submitting your resume to an organization, but not for consideration for a specific job, your resume will be scanned and the database will be used to decide whether you are suitable for any future openings. In this case, a resume that includes extra data may help your name rise higher in the electronic queue. So, rather than prepare a separate resume, you might want to simply add some terms and sections to your regular resume.

Sending Your Resume Electronically

It is likely that, at some point, a prospective employer or member of your network will ask you to e-mail them a copy of your resume. Before you blithely attach your Microsoft Word version of your resume to the e-mail, consider how formatting might get screwed up when they try to open it. Different versions of word processing programs can treat things like margins slightly differently. Default settings can be different from machine to machine, and things will certainly look different if the recipient does not have the font you used installed on his or her machine! Because resumes often have fairly complex formatting, the "as received" product can look like the Mr. Hyde version of Dr. Jekyll's resume!

Proofread your resume.

Instead of attaching a word processing file, consider converting your file to .pdf format (Adobe Acrobat is one program that can do this). The .pdf format is ideal because your resume will look exactly the same no matter what machine you send it to. Plus, it will be easy for them to print it out.

One-Paragraph and One-Page Biographies

When you contact someone for an informational interview, or when you are inquiring about a job, it is extremely helpful to give them some information about you. One excellent tool for this is a short biography.

A biography is, in essence, a version of your resume or CV that is written as plain text. It tells the story of your life and professional experience and is usually much more readable than a resume or CV. It is usually written in the third person and can contain some personal information, usually at the end.

Here is an example of a one-paragraph biography:

> Dr. Janet Thomasin is an experimental biologist in the Genomics Applications group of Immunolux Industries. She specializes in the technology of sequencing DNA using fluorescence techniques and works closely with product development groups building the next generation of Immunolux's automated genetic sequencers. A 1995 Ph.D. graduate from the University of Dayton, Dr. Thomasin is the author of 11 research publications and holds two patents. Prior to joining Immunolux in 1998, she was a postdoctoral research fellow at Smerk Pharmaceuticals in Princeville, New York.

Final Pointers, Tips, and Advice

Writing a bad resume is easy. Writing a good resume is difficult. It will take time and many drafts. Because research scientists often target several very different career paths simultaneously *it is important to have several different resumes that accent different skills*. It also goes without saying that resumes should be immaculate looking and flawless in spelling and punctuation (bad spelling is the kiss of death, so, for heaven's sake, proofread it and give it to others to read).

Here is a summary of basics and pointers:

- Support your objective/summary statement with your experiences.
- Make every word count. Use I, my, a, an, or the sparingly.
- Keep to one or two pages (one page resumes are not a rule, but stretching a resume to two is usually painfully apparent. One and a half is fine.
- No fancy fonts, strange designs, or funny colored paper unless you're applying to be an inspector at a Fruit Loop factory.
- Emphasize specific accomplishments, performance, and quantifiable results. Avoid job duties and responsibilities.
- In functional resumes, lead with the skill set that is most important for the job.
- Use action verbs in past tense.
- Be brief, positive, specific, and HONEST.
- Use numbers (30 instead of thirty).
- Don't make it crowded: 1″ margin on all sides, nice spacing between sections and experiences.
- Edit and proofread until your eyes water. One mistake is all they need....

Summary

- The purpose of a CV or a resume is to get you an interview, not a job.
- CVs and resumes are totally different documents and are not interchangeable. If you don't know which is appropriate for a particular opening, ask the prospective employer!
- Your resume, like all of your job search documents, must be immaculate and professional looking. Have friends proofread it before you send it out.

Don't get me wrong -- You have a great resume. Unfortunately, there are hundreds out there EXACTLY like you.
AJL

Six Resume Case Studies 10

Life is like a ten-speed bicycle; most of us have gears that we never use.

Linus (in "Peanuts") by Charles Schulz

You never know where a job opportunity will come from. Even during the most thorough job search, you may find yourself confronted with unforeseen opportunities in an area you hadn't considered. The key to exploiting these opportunities is to recognize them when they come, and to be able to move quickly to adapt your resume and job hunting materials to the opening. Your resume, like all of your job hunting materials, must be both attractive and professional-looking but also adaptable.

Most job seekers have never been in the position of hiring someone, so they lack the experience of sorting through a stack of resumes and picking those that looked best. This is a great exercise to do, and you'd surprise yourself with how well formed your opinions are about the look, style, and content of a resume. If you have the opportunity, I recommend that you peruse a binder of resumes to see for yourself. Collections of resumes can be found in most career centers. If you are at a scientific meeting that has a job center, you may be allowed to peruse the binders of resumes that your fellow under-employed scientists have submitted. Ask yourself which look the most appealing and which contain the best information. Are there some that are really bad? If so, why?

There are good resumes and not-so-good resumes. Fortunately, you can turn a not-so-good resume into a dynamite resume fairly easily, and you can tailor your resume to meet each opportunity as it comes. This chapter presents six "case studies": fictional people whose resumes have been restructured for the better. Each case study starts with a resume in its original form. Examine them closely and find the errors and inconsistencies. Then look at the revisions and see how the resume has been improved. For the most part, the examples are built from real people and, in some cases, the stories are completely true. As you read through these examples, keep in mind that they also illustrate some of the difficult issues that people face when they look for work, including constraints imposed by family, geography, and opportunity. In every case you will see the role that serendipity plays in today's job search.

Case Study Number 1: Janet Tangent

As an undergraduate, Janet Tangent knew from the beginning that she wanted to be a geologist. She enjoyed back-packing, skiing, sailing, and just about any other excuse to be outside. She majored in geology, had a great time as an undergrad, and decided that life could only be better in graduate school. Wary of the commitment and time that it took to get a Ph.D., Janet decided to get her Masters first. She was accepted to every program to which she applied and started the following fall at Mightybig U.

The 3 years she spent on her Masters were tough but rewarding. She produced some original research, taught students, and spent hours in the laboratory fighting with the cantankerous X-ray fluorescence spectrometer she used to analyze the rocks for her thesis. She made a number of good friends in graduate school, one to whom she became engaged in her final year. He was in his second year in the Ph.D. program.

When Janet finished she was fairly certain that she did not want to go back and invest 5 more years to get a Ph.D.; she didn't want to leave geology either. A nice opportunity opened up for her shortly before she turned in her thesis—a nearby office of a federally funded agency that carried out geological studies needed a lab technician. She applied, was hired, and for the last 2 years analyzed rocks and other samples, as well as helped out with other projects.

Things were going fine until the funding for her employer became an item on the federal chopping block. Janet was informed, in no uncertain terms, that it was very unlikely that her term appointment could be renewed. She would be out of a job in 6 months.

This wasn't an altogether unpleasant development. Janet had increasingly become frustrated with the repetition of the lab work, not to mention the paltry pay. While she and her fiancé had planned on leaving the area eventually, he still had at least 2 more years to go on his thesis. She had no idea how she was even going to support herself. So she did what most people do when they are confronted with a job crisis; she sat down and revised her resume.

Janet's first attempt (reproduced on the next page) was pretty good. It featured her work experience and research experience as a graduate student. It was fairly well designed, though a bit hard to read. However, on closer inspection, even she had to admit that it didn't say very much. The descriptions were vague and boring and made her sound like a lab mole who never saw sunlight. Janet described the things she had done so far but gave no indication of where she wanted to go. This wasn't too surprising, considering that she herself still didn't really know.

Janet Tangent

6445 Temblor Lane
Mello Park, CA 94566

TEL: (415) 555-4666
FAX: (415) 555-2199

spaz@fromage.geo.mu.edu

EDUCATION:

9/94-6/97 — **Mightybig University**, M.S. in Geological and Environmental Sciences
Thesis title: *Petrological investigations of the Cheese Wind magmatic system, Sierra Nevada, California*

8/90-5/94 — **Washington University**, BS in Geology (Civil Engineering minor), May 1994
Senior honor thesis title: *Petrology and geochemistry of the Hugh Hefner Suite, Madison 7.5" quadrangle, Virginia*

EXPERIENCE:

9/97-present — **Physical Science Aide/Technician**, Unnamed Federal Organization, CA
•Performed mineral separations, prepared samples for geochemistry, made thin sections of samples
•Digitized field maps

1/97-9/97 — **Geologist**, Jello Geotechnical, Inc.; Mello Park, California
•Quality testing of soil and liquefaction data
•Programming in dBase and use of MapInfo to create and customize earthquake databases for hazard/risk modelling

9/94-6/97 — **Lab Manager**, Mightybig University, Dept. of Geological & Environmental Sciences
•Calibrate, maintain, and operate a wavelength-dispersive x-ray fluorescence spectrometer (XRF) used for major- and trace-element analysis of rock samples
•Instruct students in sample preparation and XRF use

6/95-6/98 — **Research Assistant**, Mightybig University, Dept. of Geological & Environmental Sciences
•Field work (mapping and sampling) in Sierra Nevada, California
•Laboratory work with heavy liquid separations, polarizing microscope, scanning-electron microscope, isotope geochemistry (ion exchange columns, mass spectrometry)

9/94-6/95 — **Research Assistant**, Mightybig University, Dept. of Geological & Environmental Sciences
•Maintenance of mineral-separation, rock-crushing, and rock-sawing facilities
•Instruction of undergraduate and graduate students in use of equipment

1/96-4/96 and 10/96-6/97 — **Teaching Assistant**, Mightybig University, Dept. of Geological & Environmental Sciences
•Courses: Introductory Geology, Volcanology, Igneous and Metamorphic Petrology
•Prepared laboratory exercises, taught lab sections, graded exercises and exams

SKILLS:

•Extensive experience with Macintosh and Microsoft Windows-based software
•Experience with UNIX, DOS, and RSX-11M operating systems
•Programming ability in Turbo Pascal, FORTRAN, and dBase
•Presentation of research results at national meetings in both oral and poster formats
•Four years of French, two years of German

Janet the Geotechnical Engineer

On the advice of a friend, she made an appointment with a career counselor at the Career Planning and Placement Center at Mightybig U. He guided her through several self-assessment exercises and got her to arrange several informational interviews. After doing all this, and talking to her friends, she came up with two possible career paths to occupy her for the next 2 years of her life.

The first was the field of environmental or geotechnical engineering. She had some experience in this field, having worked freelance for a geotechnical firm while she was finishing up her Masters. She liked the work, especially the chance to do much of it outside. She also liked that it called upon the technical skills she developed in her Masters. After conducting several informational interviews and reading a book about the field of geotechnical engineering, she revised her resume to target this particular field (see facing page).

As you can see, this resume is much stronger than the original. For starters, it is clear that Janet has a specific objective in mind. Then, using what she learned about the field of geotechnical engineering from the informational interviews, she identified particular skills that were valuable. One of the best aspects of her resume is that she demonstrates her record of performance with numbers. The layout is clean, it is easy to read, and it has a professional look.

Janet the Freelance Desktop Publisher

In the process of putting together her geotechnical resume, Janet talked to the mother of a friend of hers who, for many years, has been a freelance writer and desktop publisher. Janet discovered that this type of work could pay as well as an entry-level job in a geotechnical firm, but had much more flexibility. Instead of a daily commute she could work from home on a computer she already owned.

Janet learned from her source that one way to get established is to join a temp agency that supplies jobs to technical writers. The quality of the assignments and the compensation depended on experience. Hence, she changed her resume again, bringing her technical writing, graphic design, and document production experience to the top (see page 110).

This resume, like the geotechnical resume, demonstrates her experience in quantitative terms. In addition, the resume lists all of the computer programs with which Janet is familiar. Most important, Janet rephrased the descriptions of her past experience to better describe her technical writing qualifications. It is clear that each resume is tailored to the specific industry that she has in mind.

Janet Tangent

6445 Temblor Lane
Mello Park, CA 94566
E-mail: spaz@fromage.geo.mu.edu
Tel: (415) 555-4666

OBJECTIVE:

Challenging position as a geotechnical or environmental engineer utilizing proven analytical, computer, and communication skills

EDUCATION:

Mightybig University, Bigville, California
M.S. in Geological and Environmental Sciences **1997**

Washington University, St. Louis, Missouri
B.S. in Geology – Magna cum Laude (Minor in Civil Engineering) **1994**

HONORS AND AWARDS:

Outstanding Teaching Assistant Award, Mightybig University **1996**
Arthur Buddington Award, Department of Geology, Washington University **1994**
National Merit Scholarship, Semi-finalist **1990**

TECHNICAL EXPERIENCE:

Physical Science Aide/Laboratory Technician **1998 - present**
Unnamed Federal Organization, Mello Park, California

- Assisted in chemical analysis of geological samples using optical microscopy and ICP-mass spectrometer
- Digitized and modified topographic, geologic, and land-use maps
- Assisted in preparation of 3 published articles and 2 internal reports

Technical Consultant **1997**
Jello Geotechnical, Inc., Mello Park, California

- Collected, measured, and evaluated engineering properties of soils
- Developed and programmed custom earthquake databases for seismic hazard modeling of properties using dBase, MapInfo, Microsoft Excel, and GIS software
- Wrote and presented risk assessment reports to clients

Lab Manager **1995 - 1999**
Department of Geological & Environmental Sciences, Mightybig University, Bigville, California

- Calibrated, operated, maintained, and repaired X-ray Fluorescence Spectrometer
- Developed and administered billing and operating procedures that cut laboratory costs by 60%
- Trained 23 users in safe operating procedures

Research Assistant **1994-1997**
Department of Geological & Environmental Sciences, Mightybig University, Bigville, California

- Organized and executed original scientific research on volcanic rocks from eastern California. Work included: geologic mapping, sample collection and characterization, chemical analysis, and radiometric age dating
- Developed novel technique for mineral separation and characterization
- Wrote 3 research papers (published/in press), presented 5 papers at national meetings, and led 10 seminars

ADDITIONAL SKILLS:

Foreign Languages *Speaking/writing/reading proficiency in French and German
Computer *Analytical programming in FORTRAN, Turbo Pascal, and dBase
*Experience with UNIX, DOS, Apple, and RSX-11M operating systems

Janet Tangent

6445 Temblor Lane
Mello Park, CA 94566
E-mail: spaz@fromage.geo.mu.edu
Tel: (415) 555-4666

OBJECTIVE: Freelance or part-time position as a technical writer/graphic designer utilizing extensive computer experience and effective communication skills

WORK EXPERIENCE:

Physical Science Aid/Laboratory Technician **1998 - present**
Unnamed Federal Organization, Mello Park, California
- Digitized and modified topographic, geologic and land-use maps
- Assisted in the preparation of 3 published articles and 2 internal reports
- Drafted and modified scientific figures for publication

Technical Consultant **1997**
Jello Geotechnical, Inc., Mello Park, California
- Wrote and presented publication-quality risk assessment reports to clients
- Designed and produced technical and schematic graphics using Adobe Illustrator, Adobe Photoshop, MacDraw Pro, and Superpaint
- Developed and programmed custom earthquake databases for seismic hazard modeling of properties using dBase, MapInfo, Microsoft Excel, and GIS software
- Collected, measured, and evaluated engineering properties of soils

Teaching Assistant **1995, 1997**
Department of Geological & Environmental Sciences, Mightybig University, Bigville, California
- Designed, prepared, and taught laboratory exercises to 30 students
- Prepared 85-page laboratory exercise book with 40 original figures and diagrams
- Developed exam materials and graded course work with professor

Research Assistant **1994 - 1997**
Department of Geological & Environmental Sciences, Mightybig University, Bigville, California
- Organized and executed original scientific research on volcanic rocks from eastern California
- Wrote 3 research papers (published/in press), presented 5 papers at national meetings, and led 10 seminars
- Designed novel laboratory device using Microsoft CADCAM

COMPUTER SKILLS:
- Extensive computer graphics design experience using Adobe Illustrator, Adobe Photoshop, MacDraw Pro, Claris SuperPaint, and Microsoft CADCAM
- Advanced programming ability in FORTRAN, Turbo Pascal, and dBase
- Advanced word processing with Quark Xpress, Microsoft Word, and MacWrite
- Extensive experience with Macintosh, Windows, UNIX, DOS, Apple, and RSX-11M operating systems

EDUCATION

Mightybig University, Bigville, California
M.S. in Geological and Environmental Sciences **1999**

Washington University, St. Louis, Missouri
B.S. in Geology – Magna cum Laude (Minor in Civil Engineering) **1994**

Case Study Number 2: Dr. Sigmund Becker

Long before he finished his Ph.D., Sigmund Becker (Sig to his friends) knew that he wanted to do something different than laboratory-based research. His transition took place over 3 years-from the final year of his Ph.D. through his 2 years as a postdoc.

Sig had graduated from a prestigious, small liberal arts college where he had great mentorship and the opportunity to conduct research during his junior and senior years. By chance, his favorite professor was a biochemist. Miraculously, Sig's love of science survived graduate school. Sig found himself in a group of 28 graduate students, 11 postdocs, four research associates and one hyper-kinetic, sleep-deprived principal investigator. Sig got hooked on nuclear magnetic resonance spectroscopy (NMR) and studied the structure of carbohydrates.

Sig was sure that he wanted to do something new by the time he was nearing the end of his Ph.D. While the prospects of 3-5 years as a postdoc followed by a steep path toward gainful employment in research left him a little queasy, Sig was mostly motivated by a desire to work on a broader range of issues than his narrow field of specialization.

Sig liked all sorts of science and read avidly. He first thought about science journalism and explored the career by talking to a friend of a friend who had gone that route. He thought about science museums and science education. All these careers seemed to combine his broad interest in science with his delight in working with people.

Sig thought long and hard about whether to do a postdoc at all. In the end, he decided to apply for a postdoctoral fellowship at a famous biology research institute in the San Francisco area. For him it was a question of location. The Bay Area had a number of outstanding museums, great opportunities for science writers, and weather that couldn't be beat! Sig figured that he could spend his first few years as a postdoc networking in the area and building contacts to help in his eventual transition.

Once in the Bay Area, however, Sig got bitten by the start-up bug. In his first year, Sig saw three members of the technical staff leave to start their own company. One of them was Sig's office mate. Sig kept in touch and got a daily report on the thrills and chills of his friend's experience. After 3 months, the new start-up was up against a huge problem nobody had anticipated: a lawsuit over intellectual property.

Sig the Patent Lawyer

As Sig heard more details about the predicament his friend's company was in, he became more and more interested in the field of patent law. It turned out that one of the lawyers hired to help bail the company out of trouble was, herself, a Ph.D. biochemist who had done NMR spectroscopy. Small world!

Sig met with Jeannie, the patent lawyer, the following week. Jeannie told him all about her transition, and how she landed a job as a patent inspector for a large law firm, and how the law firm paid for her entire law degree at a prestigious university in the Bay Area. Sig handed her his resume and asked for suggestions.

Jeannie looked the resume over carefully. "You know," she said, looking up at Sig, "I'm sure you've done a whole lot more than you've got listed here." Jeannie proceeded to grill Sig about the research areas he was familiar with. "Do you know anything about assays for amino acids?" she asked him.

"I do", replied Sig, "but I really didn't publish in that..."

"Doesn't matter!" exclaimed Jeannie, "what legal departments are looking for is breadth, not depth! You need to show the breadth of your technical background and all the commercially important areas that you are familiar with."

Jeannie told Sig to sit down and write out all the areas of science and technology he was familiar with, along with any major technological applications from those fields. Sig dutifully complied, not only because it seemed like a good way of getting organized and understanding the scope of his experience, but also because Jeannie was really cute!

Sig's list was over two pages long and took him a weekend to assemble. Sig had worked on proteins, carbohydrates, enzymes, DNA, had used NMR, EPR, and about a dozen other analytical techniques. He had done computational modeling as well as lots of wet chemistry. He had dealt with fluorescent tagging techniques, assays, separations, and purifications. The list was dizzying!

The following week Sig took his list back to Jeannie and the two of them sorted through the relevant technological issues. Jeannie told him which areas were hot fields for intellectual property and together they crafted his new resume. Jeannie encouraged Sig to include some of the "extracurricular" activities that he had been engaged in over the last few years, especially two small consulting projects he undertook with another postdoc. Sig also had some leadership experience as a postdoc and in graduate school, and Jeannie encouraged him to include this information as well.

"Including the leadership stuff makes you look more well-rounded," Jeannie explained, "and they won't think of you as just a geek."

Sigmund's revised resume (see pages 114 and 115) looked much better- and was quite a bit longer.

Through Jeannie, Sigmund did several informational interviews with some of the leading intellectual property law firms in the Bay Area. One firm, Dewey, Chetham and Howe, offered him a position as a patent agent trainee. Sig finished his postdoc and began his new life as a patent agent. He also began dating Jeannie!

Sigmund F. Becker

12322 Poppy Ave.
Apt. 2
Santa Opaqua, California, 94555
650-555-8787

Department of Biochemistry
The Loadstone Institute
2000 25th Ave
San Francisco, CA 95677
415-555-9000

Education

Ph.D. in Biochemistry, 1998, University of Minnesota

A. B. in Biology (dual major in Chemistry), 1990, Wilhelms College

Research Experience

4/98 - present — Senior Research Associate:, Dr. Haywood Jubuzzoff, Chairman, Department of Biochemistry, The Loadstone Institute, San Francisco, CA. Determining the solution structure of Testosterone-Related Receptor-4 protein - DNA complex using multidimensional, heteronuclear NMR spectroscopy. Developing methodology to expedite structure determination by NMR.

9/90 - 4/98 — Graduate Student Research Assistant: Prof. Marge Ennoverra, Department of Biology, University of Minnesota. Studied the solution conformation of oligosaccharides with multi-dimensional NMR spectroscopy.

1/93 - 6/94 — Senior Instructor, University of Minnesota Center for Health Sciences. Supervised laboratory instructors and taught lecture sections. Developed curricula.

6/91, 6/92 — Instructor, University of Minnesota Department of Biology. Taught Biochemistry laboratory sections.

9/89 - 6/90 — Research Technician; Prof. Slo Lerner, Center for Bioanalytical Research, Bradley AL. Developed HPLC methods for the determination of amino acids and short polypeptides at sub-nanomolar levels.

Awards

Loadstone Institute Society of Fellows 9th Annual Symposium Poster Award (6/99)
NIH/NIGMS Postdoctoral Fellowship (4/98-4/00)
Department of Education Fellowship (9/91 - 9/93)
Merck Award (1990)
National Merit Scholarship Finalist (1985)

Society Affiliations

American Association for the Promulgation of Science
American Biochemical Society
Iota Nada Pi

Sigmund F. Becker

12322 Poppy Ave.
Apt. 2
Santa Opaqua, CA, 94555
Tel: (650) 555-8787

Department of Biochemistry
The Loadstone Institute
2000 25th Ave.
San Francisco, CA 95677
Tel: (415) 555-9000

Profile

- Technical areas of expertise: biophysics, chemistry, molecular biology, protein biochemistry
- Specialized expertise: structural biology, molecular modeling, genomics, bioinformatics, proteins, enzymes, molecular recognition, assay techniques, fluorescence tagging, carbohydrates, magnetic resonance imaging (MRI), microfluidics, nuclear magnetic resonance spectroscopy (NMR)

Education

University of Minnesota, St. Paul, Minnesota — **1998**
Ph.D. in Biochemistry

Wilhelms College, Dearborn, Massachusetts — **1990**
A.B. in Biology (dual major in Chemistry)

Technical Experience

Senior Research Associate — **1998–present**
Department of Biochemistry, The Loadstone Institute, San Francisco, California

- developed new modeling method that improved efficiency of biomolecular structure determination
- collected and analyzed structural data on protein-DNA complex related to Alzheimer's disease using nuclear magnetic resonance spectroscopy (NMR)
- built and modified analysis cell for commercial flow cytometer. Improved analysis precision and speed by 45%

Research Assistant — **1990–1998**
Department of Biology, University of Minnesota, St. Paul, Minnesota

- solved the 3-dimensional structure of a transcription factor protein important in tissue cell development using NMR and data and molecular force fields in a simulated annealing protocol
- developed new HPLC method for separation of hyaluronan oligosaccharides

Research Technician — **1989–1990**
Center for Bioanalytical Research, Bradley, Alabama

- assisted in developing assay for amino acids with 300% increase in sensitivity above existing methods
- studied kinetics of fluorescent tag reactions for labeling amino acids

Research Assistant — **1989–1990**
Department of Biology, Wilhelms College, Dearborn, Massachusetts

- developed enzymatic assay for nitrate for use in medical applications
- studied kinetics of enzyme oxalate oxidase for diagnosis and treatment of kidney stones

Consulting Experience

- Evaluated technical feasibility of small biotech company's diagnostic equipment idea (infrared microimager) for investment group associated with University of California at San Francisco's Technology Forum (1999)

- Summarized information on viral diseases for Golworth & Co., LLP, to help evaluate potential utility of licensed compound for treatment of viral infections in animals (1998)

Teaching Experience

Senior Instructor **1991–1994**
University of Minnesota Center for Health Sciences, St. Paul, Minnesota
- developed laboratory-based curriculum for undergraduate class in Biochemistry
- taught weekly lecture sections to 90 students
- supervised 8 teaching assistants

Teaching Assistant Trainer/Teaching Assistant **1988–1990**
Department of Biology, Wilhelms College, Dearborn, Massachusetts
- trained undergraduate instructors for laboratory-based undergraduate class (Laboratory Methods in Biology, Bio 203)
- wrote, administered, and graded weekly quizzes and midterm examinations
- led discussions on teaching methods, professional behavior, and college policies

Leadership and Communication Experience

President of The Loadstone Institute Postdoctoral Society **1999–present**
- lead a 12-member committee in organizing lecture series, symposia, and fund-raising events
- initiated and led programs to improve postdoctoral fellows contact with biotech industry

University of Minnesota Student-Faculty Liaison Committee **1995–1998**
- one of 6 representatives to campus-wide committee representing 1100 graduate students
- arbitrated disputes between students and faculty and assisted in mediating conflicts regarding tuition support, housing, leave and benefits, and intellectual property

Awards

National Institute of Health Postdoctoral Fellowship	1999–present
Department of Education Graduate School Fellowship	1991–1994
National Science Foundation Summer Biochemistry Fellowship	1988
Merck Award (Wilhelms College)	1988

Case Study Number 3: Dr. William M. S. Dos

Dr. William (Bill) M. S. Dos has been a geophysicist for the National Ocean Water Agency (NOWA) for the last 5 years. He was considered a young hot-shot when he first got his job in 1995, and his performance since then has been exemplary. Bill had always assumed he would be working for NOWA for much of his career, rising in the ranks of the organization, and perhaps owning one of those nifty sailor uniforms that NOWA "officers" get to wear!

However, in only a year things changed dramatically for NOWA in general and Bill's research group in particular. In the last few months there had been very specific talk of cuts in research, although no formal actions were taken. Bill and the rest of his colleagues were worried. Bill was particularly concerned because he was 3 years into a 20-year mortgage on a house in a nice part of town and his wife was pregnant. The possibility of losing his well-paying job at NOWA left him anxious and distracted. What should he do?

Bill started by updating his CV. Not only did he revise his publication list but he included a description of some of the other professional activities he had engaged in over the past few years. It looked pretty good (see pages 118-119).

However, Bill was at a total loss about what to do with it. Should he start applying for faculty positions? He had considered that route before but felt that he lacked sufficient teaching experience and had a career that was more directed to research. There were occasional openings for research faculty positions around the country but none recently and Bill was also fairly certain that relocating beyond the east coast was out of the question.

Like many scientists, Bill was a good public speaker, a concise, careful, and experienced writer, and had an excellent background and ability with quantitative analysis and computers. He initially considered computer programming as an obvious career target but realized that, while he was eminently qualified, he preferred to work with people rather than machines. He also found the prospect of dealing with incessant demands from software users a bit nauseating. In fact, Bill had developed a package of data reduction software as a graduate student and found the "tech support" to be a huge time sink and source of frustration.

So Bill began doing some career planning research on his own. He got a membership to a local career planning and placement center and, after an interview with a career counselor and some self-assessment exercises, began a sincere effort to research alternative careers.

Bill Explores Other Careers

Bill looked back into his past experiences to find out what he enjoyed doing most. One of the most interesting and enjoyable experiences he had was as an expert witness in a lawsuit. This called upon good communication and logical skills in a setting that he found very exciting. The experience had also introduced him to some broader applications of his particular technical skill in the fields of oil exploration and airborne inertial navigation. During this project he had met three people, a physicist from Excon Oil Company, a physicist from a defense contracting firm called Corral, and a former chemist (now lawyer) who worked on the case. Bill resolved to contact these people and inquire about opportunities working on the broader applications of his research in geodesy and geophysics.

However, Bill was encouraged to consider multiple pathways and thus returned to the possibility of working as a computer programmer. After meeting with a programmer from Moon Microsystems he had to admit that his preconceptions about the life of a computer programmer were badly in error. Rather than the isolated, nerdy experience he imagined, most of the programmers he met worked in tight teams with a great deal of interaction with other teams. They were younger and more "normal" than he had expected! So maybe a life as a programmer wouldn't be too bad. After all, it would pay more that what he was earning currently.

Bill had one other unrequited interest in his life: music. Bill sang in several choirs, played the piano, and had always assumed that this would be relegated to a hobby for the rest of his life. However, in the process of talking to programmers, he heard about a small start-up company that specialized in acoustic design and music. This outfit had one product already, software capable of analyzing the acoustics of performance halls. Bill heard that they were in the process of growing and thought that his dual background in science and the arts might be a good match.

Bill produced two resumes, one for the technical people at Excon and Corral (pages 120-121) and another for the music company (page 123). Both use similar material but one is two pages, emphasizing Bill's technical experience while the other is only one page. One job is clearly an "entry-level" position.

As you will see, there is no such thing as a "perfect" resume. People write resumes differently. The critical step to preparing a winning resume is to research the job thoroughly. Try to match as many of the company's needs as possible in the resume and cover letter.

Bill Applies for a Faculty Job

A potential opportunity arose right under Bill's nose while he was exploring his career options outside of research. In the latest issue of GLOBUS, the Transactions of the American Metaphysical Association, he read the following advertisement:

Laughlin College

Ocean Scientist

The Department of Geology seeks an assistant professor, tenure-track, to begin August of this year. The successful candidate will teach oceanography, geophysics, and a course of his or her specialty. A Ph.D. in the geosciences, teaching experience, and an ongoing program of research are required. Please send letter of application, vita, graduate and undergraduate transcripts, statements of teaching and research interests, and three letters of recommendation to: Bill Baker, Chair, Department of Geology, Laughlin College, Neptune City, RI 10023. Application deadline: March 30, 2000. Laughlin College is a highly selective liberal arts college that has demonstrated its commitment to equal opportunity and the promotion of cultural pluralism. Women and members of minority groups are particularly encouraged to apply. EOE/AA.

Curriculum Vitae
William M.S. Dos
(revised December 1, 1998)

Date of Birth:	July 4, 1961	Social Security No:	765-66-7676
Place of Birth:	Providence, Rhode Island	Citizenship:	United States

Education

Degrees Earned:

October 1993	Ph.D., Marine Geophysics, Massachusetts Institute of Learning, Massachusetts Dissertation: *Marine Geophysical Studies of Mid-Ocean Ridges*
May 1989	M. A., Geological Sciences, Massachusetts Institute of Learning, Massachusetts
May 1987	B. S. Geological Sciences, Beans University, Houston, Texas

Other relevant studies:

July 1987	Ecole du Monde: Geophysique Interne et l'Espace, Centre National d'Etudes Spatiales, Nice, France
1982-83	Undergraduate studies at Puny College, Puny, NY (transferred to Beans in 1980)

Professional Experience

Positions Held:

1995-	Geophysicist, Geophysical Working Group, National Aquatic Service, Nat'l Ocean Water Agency, U.S. Department of Commerce, Silver Spring, MD.
1993-95	Postgraduate Researcher, Institute of Geoplanetary Science, Scripts Institution of Oceanography, University of California, La Jolla, CA
1987-93	Graduate Research Assistant, Lamont-Doughy Geological Observatory of Massachusetts Institute of Learning, Massachusetts
Spring 1990	Teaching Assistant in Structural Geology, Massachusetts Institute of Learning, Massachusetts

Field and Sea Experience:

November 1994	R/V *George Washington* cruise Tunes-05: Co-Chief Scientist, Mid-Atlantic Ridge
July 1993	R/V Great Wave cruise GW-9009: Co-Chief Scientist, Cape Verde Islands Guyots ODP Site Survey Augmentation
May 1991	R/V George Washington cruise RNDB-II: responsible for acquisition and reduction of gravity data for Mid-Atlantic Ridge Geophysical and Coring Survey
September 1988	R/V Washington Irving cruise RC-2610: responsible for acquisition of data for thesis research including gravity, Seabeam, and dredging operations, AT&T Cable Survey/ Mid-Atlantic Ridge

Summer 1986	Summer Field Camp, Red Lodge, MT, Beans University
Spring 1986	Field Mapping, Mojave Desert, CA, Beans University

Professional Society Memberships

1994	Sigma Xi
1993	Society of Exploration Geophysicists
1988	American Metaphysical Association

Honors and Awards

1998	Quality Step Increase for Outstanding Performance, NOWA
1997	Cash Award for Outstanding Performance, NOWA
1994-96	Cecil and Ida Brown Foundation Scholar, Institute of Geoplanetary Science, Scripts Institution of Oceanography
1987-93	Faculty Fellowship, Massachusetts Institute of Learning
1986	Summer Internship, Lamont-Doughy Geological Observatory
1985-88	Peck Foundation Merit Award, Beans University

Service to the Scientific and Educational Community

Since 1991, co-author and distributor of the PointerCalc software system. This is a collection of C language tools, data, and on-line help for processing and displaying vector-based data sets. It is distributed free of charge over the Internet, and is used by more than 5000 scientists on every continent (including Antarctica) and on board ships and aircraft.

Member, Committee on Education and Employment, American Metaphysical Association (July 1997 to June 1999).

Scientific Advisor to the International Hydrologic Organization and the Inter-governmental Oceanographic Commission of UNESCO for GEBCO, the General Bathymetric Chart of the Oceans. (Officially since June, 1997; unofficially serving since May, 1996).

Pro bono consultant to the National Geographic Society for the revision of its World Physical Map. Directed artist Bill Bother to paint sea floor relief in accordance with new results from satellite altimetry. Map was published with acknowledgment of this contribution in the February 1998 issue of the *National Geographic* magazine.

Expert witness in Muckraker v. BigOil, a patent infringement lawsuit concerning a patented process for making geoid maps from satellite altimetry for oil exploration purposes (Summer and Fall, 1997).

Peer Reviewer of research proposals submitted to the National Science Foundation and the National Aeronautics and Space Administration.

Peer reviewer of research articles submitted to the *Journal of Metaphysical Research*, *Earth and Lunatary Science Letters*, *Metaphysical Journal International*, *Metaphysical Research Letters*, and *Teutonics*.

Publications

(Insert here a list of 33 peer-reviewed publications (16 as first author), 15 technical reports/data packages, and abstracts since 1996.)

William M. S. Dos

633 Bent Branch Road
Chevy Chase, Maryland 20840
Tel: (301) 555-4001
E-mail: wdos@geo.nowa.gov

Objective: Challenging position developing gravity-based guidance and exploration techniques for navigation and oil exploration in a team-oriented environment that utilizes my proven expertise in computational geophysics.

Education: Ph.D., Marine Geophysics, Massachusetts Institute of Learning (1995)

M.A., Geological Sciences, Massachusetts Institute of Learning (1989)

B.S., Geological Sciences, Beans University (Cum Laude) (1987)

Technical Experience:

1995 to present — **Research Geophysicist:** Geophysical Working Group, National Aquatic Service, National Ocean Water Agency (NOWA), Chevy Chase, Maryland

Developed novel computational techniques for processing satellite-based gravity data used in geodesy, navigation, and weather prediction. Initiated and oversaw $500,000 research program in satellite-based ocean bathymetry. Coordinated research efforts among 5 groups at NOWA. Wrote 8 papers published in scientific journals, and presented 15 papers at national and international meetings. Supervised 2 assistant researchers.

1993-95 — **Cecil and Ida Brown Foundation Scholar:** Institute of Geoplanetary Science, Scripts Institution of Oceanography, University of California, La Jolla, California

Analyzed, processed, and interpreted satellite-based and ship-based geophysical data. Produced quantitative gravity data sets for use by scientists and industry. Collaborated with international team of scientists, published 4 papers in scientific journals and presented 8 papers at national and international meetings.

1987-93 — **Research Assistant:** Lamont-Doughy Geological Observatory of Massachusetts Institute of Learning, Our Fair City, Massachusetts

Developed, co-wrote, and distributed the PointerCalc software package, a collection of C language tools, data, and on-line help for processing and displaying vector-based data sets. It is distributed free of charge over the Internet, and is used by more than 5000 scientists on every continent (including Antarctica) and onboard ships and aircraft. Taught upper-level courses in marine geophysics.

Related Consulting Experience:

American Metaphysical Association: Invited member of the Committee on Education and Employment. Advised and oversaw secondary and primary education programs and human resource services for 31,000-member scientific society. (1998 - present)

International Hydrologic Organization and the Inter-governmental Oceanographic Commission of UNESCO: Advised United Nations agency on technical veracity and production of the General Bathymetric Chart of the Oceans (GEBCO), an international project providing navigational information to world governments. (1997 - present)

National Geographic Society: Provided technical information and advised design of the revision of its World Physical Map. Directed artist Bill Bother to paint sea floor relief in accordance with new results from satellite altimetry. Map was published in the February 1994 issue. (1997)

Dewey, Chetham and Howe: Expert witness in Muckraker v. BigOil, a patent infringement lawsuit concerning a patented process for making geoid maps from satellite altimetry for oil exploration purposes. (1997)

Honors and Awards:

1998	Quality Step Increase for Outstanding Performance, NOWA
1997	Cash Award for Outstanding Performance, NOWA
1994-96	Cecil and Ida Brown Foundation Scholar, Institute for Geoplanetary Science, Scripts Institution of Oceanography
1987-93	Faculty Fellowship, Massachusetts Institute of Learning
1986	Summer Internship, Lamont-Doughy Geological Observatory
1985-88	Peck Foundation Merit Award, Beans University

Selected Publications:

(Insert here a list of 6 publications most relevant to the interests of the employer.)

Up until that point, Bill had not seriously considered a teaching career. His graduate school experience and his subsequent employment had simply pointed him in another direction. In fact, Bill had begun teaching a course at a nearby university last year because his current job did not give him any opportunity to teach students, something he did enjoy doing.

Bill started by researching the job and the school. Laughlin was an excellent, small liberal arts college and its Geology Department consisted of five faculty members, one of whom had just been denied tenure. Bill called the department chair and discussed the needs of the school. They wanted someone who was both an excellent teacher AND who could incorporate undergraduates into his or her research program. The undergraduate research part was important: according to the chairman, every senior in the department is required to do a thesis, and the last professor they hired was denied tenure mainly on the grounds that his research (modeling ocean circulation) was too complex for undergraduates. Bill also called the professor who was denied tenure and asked him about the department and the job.

"That line about denying me tenure because I didn't have undergraduates working with me is %$#@!!!" the jilted academic responded when Bill told him what he'd heard from the department chair.

"The fact is," the professor continued, " I didn't get along with the chairman who, in my opinion, is a fatuous, insecure, pompous jerk who hasn't published anything since the '70s and was intimidated by my research. Be careful: if you tick this guy off, you're toast. My advice to you is: emphasize your teaching and the opportunities for undergraduate research, and good luck."

Bill thanked the guy and then sat down to compose his materials. He revised his resume, splitting his professional experience into Teaching and Research sections. He tried to emphasize his teaching record as much as possible. In his teaching statement he listed the courses he felt prepared to teach and discussed his philosophy of teaching, which emphasized applied problem solving and project-oriented assignments. He also emphasized the possibility of teaching environmental science courses, as he knew this was an area of growing interest among undergraduates at this school.

However, it was his research statement that he really overhauled. He started by altering the scope of the research so that it was clear that his goals were feasible for a lone professor at a small school. He emphasized that, if hired by Laughlin, he would retain access to the facilities and resources, and perhaps even the funding, from NOWA. Most importantly, he stressed the role that undergraduates could play in the research program. He made a point of emphasizing that NOWA provided undergraduates with opportunities to do a semester of research at sea (a VERY popular program at Bill's undergraduate school) and that undergraduates could conduct a wide variety of collaborative research with him and other professors at Laughlin.

He emphasized in his cover letter that he had a great interest in undergraduate education, research experience that complemented the department, and was eager to incorporate undergraduates into his research program. He also mentioned the possibility of retaining NOWA funding as a professor. He made it a point NOT to include a thick stack of reprints.

William M. S. Dos
633 Bent Branch Road
Chevy Chase, Maryland 20840
Tel: (301) 555-4001
E-mail: wdos@geo.nowa.gov

Technical/Marketing Experience:

1993 to present **Research Geophysicist:** Geophysical Working Group, National Aquatic Service, National Ocean Water Agency, Chevy Chase, Maryland (1995 - present), Institute of Geoplanetary Science, Scripts Institution of Oceanography, University of California, La Jolla, California (1993-95)

Developed novel computational techniques in C and C++ for processing satellite-based gravity data from a variety of instruments used in geodesy, navigation, and weather prediction. Initiated and oversaw $500,000 research program in satellite-based ocean bathymetry. Coordinated research efforts among 5 groups at NOWA. Wrote 12 papers published in scientific journals, presented 23 papers at national and international meetings. Supervised 2 assistant researchers.

1987-93 **Research Assistant:** Lamont-Doughy Geological Observatory of Massachusetts Institute of Learning, Our Fair City, Massachusetts

Developed, co-wrote, marketed, and distributed the PointerCalc software package, a collection of C language tools, data, and on-line help for processing and displaying vector-based data sets. It is distributed free of charge over the Internet, and is used by more than 5000 scientists on every continent (including Antarctica) and onboard ships and aircraft. Developed course materials and taught upper-level courses in Marine Geophysics.

Music Experience:

National Cathedral Chorale: Assistant choir master (1996-present). Lead bass vocalist (1998). As assistant choir master, organized and ran rehearsals, auditioned new singers, and organized spring concert series. As lead bass vocalist, sung variety of solos from Mozart, Puccini, and Philip Glass.

Columbia University Choir: Lead bass vocalist (1991-92). Pianist (1988-92). Performed modern and classical choral music with the Columbia University Symphony at Davis Symphony Hall and Carnegie Hall. As pianist, led rehearsals and accompanied symphony.

Education:

Ph.D., Marine Geophysics, Massachusetts Institute of Learning, Our Fair, Massachussetts (1994)
M. A., Geological Sciences, Massachusetts Institute of Learning, Our Fair, Massachussetts (1989)
B. S. Geological Sciences (Minor: Music), Beans University, Houston, Texas (1987)

Computer Skills:

Extensive programming experience in UNIX environments using C, C++, AWK; specialized in IO interface with serial interface devices
Programming experience with Macintosh-based systems incorporating MacTools
Experience with MS Windows and OS/2 operating systems
Expert user of Microsoft Word, Adobe Illustrator, Adobe Photoshop, and Microsoft Excel

CURRICULUM VITAE
WILLIAM M. S. DOS

Geophysics Working Group
National Aquatic Service
633 Bent Branch Road
Chevy Chase, Maryland 20840
Tel: (301) 555-4312
E-mail: wdos@geo.nowa.gov

Personal

Citizenship: United States of America

Education

Ph.D., Department of Geophysical Sciences, Massachusetts Institute of Learning **1995**
Thesis: Marine Geophysical Studies of Mid-Ocean Ridges
Principal Advisor: Professor Charles A. Tuna

M. A., Department of Geophysical Sciences, Massachusetts Institute of Learning **1989**

B. S., Department of Geological Sciences, Beans University **1987**
(Magna Cum Laude, with Geological Engineering Certificate)

Teaching Experience

Lecturer **1999–present**
Department of Earth and Space Sciences, John Hopmeister University, Baltimore, Maryland
Teach 3-unit course, *Potential Field Theory*, for upper-division undergraduate and graduate students in Geophysics, Physics, and Astronomy. Design and administer weekly problem sets and computer laboratory exercises. Current enrollment: 13 students.

Teaching Assistant **1988–1993**
Department of Geophysical Sciences, Massachusetts Institute of Learning, Our Fair City, Massachusetts
Sole teaching assistant for *Introduction to Oceanography* (instructor: Charles Tuna, enrollment: 70) Fall 1988-Spring 1989. Teaching assistant for *Geophysical Methods in Ocean Science* (instructor: Wallace Whale, enrollment: 10) Fall 1989-Spring 1992. Organized and taught *Earth Science Applications of the PointerCalc Computer Program* and *Computer Applications for Geophysicists* (enrollment: 7).

Research Experience

Research Geophysicist **1995–present**
Geophysical Working Group, National Aquatic Service, National Ocean Water Agency (NOWA), Chevy Chase, Maryland
Developed novel computational techniques for processing satellite-based gravity data and used high-resolution bathymetric data to interpret tectonic and gravity features of mid-ocean ridges. Designed and oversaw shipboard geophysical surveys on NOWA cruises 1233 and 1235. Initiated and oversaw $500,000 research program in satellite-based ocean bathymetry. Coordinated research efforts between 5 groups at NOWA. Supervised 2 assistant researchers.

Cecil and Ida Brown Foundation Scholar **1993–1995**
Institute of Geoplanetary Science, Scripts Institution of Oceanography, University of California, La Jolla, California
Analyzed, processed, and interpreted satellite-based and ship-based geophysical data. Produced quantitative gravity data sets for use by scientists and industry. Collected and processed on-board gravity data from R/V *George Washington* cruise Tunes-05 and R/V *Great Wave* cruise GW-9009.

Research Assistant **1987–1993**

Lamont-Doughy Geological Observatory of Massachusetts Institute of Learning, Our Fair City, Massachusetts

Investigated mid-ocean ridge tectonics using ship-based and satellite-based geophysical data. Developed, co-wrote, and distributed the PointerCalc software package, a collection of C language tools, data, and on-line help for processing and displaying vector-based data sets. Acquired and processed on-board gravity data from R/V *George Washington* cruise RNDB-II: Mid-Atlantic Ridge Geophysical and Coring Survey. Acquired and processed gravity, Seabeam and dredging data for R/V *Washington Irving* cruise RC-2610: AT&T Cable Survey/ Mid-Atlantic Ridge.

Professional Activities

Member, Committee on Education and Employment, American Metaphysical Association

In collaboration with Professor Bashful Sampson (St. Paul University) organized panel discussions at the Fall 1997 AMA meeting on alternative career paths for Ph.Ds. As a committee member, advise AMA on education projects and spending and allocate resources for career development activities. (1996 - present)

Scientific Consultant, International Hydrologic Organization and the Inter-governmental Oceanographic Commission of UNESCO

Advised United Nations agency on technical veracity and production of the General Bathymetric Chart of the Oceans (GEBCO), an international project providing navigational information to world governments. (1997 - present)

Scientific Consultant, National Geographic Society

Provided technical information and advised design of the revision of its World Physical Map. Directed artist Bill Bother to paint sea floor relief in accordance with new results from satellite altimetry. Map published in the February 1994 issue of *National Geographic* magazine. (1997)

Expert Witness, Dewey, Chetham and Howe

Provided expert testimony in Muckraker v. BigOil, a patent infringement lawsuit concerning a patented process for making geoid maps from satellite altimetry for oil exploration purposes. (1997)

Honors and Awards

1998	Quality Step Increase for Outstanding Performance, NOWA
1997	Cash Award for Outstanding Performance, NOWA
1994-96	Cecil and Ida Brown Foundation Scholar, Institute of Geoplanetary Science, Scripts Institution of Oceanography
1987-93	Faculty Fellowship, Massachusetts Institute of Learning
1986	Summer Internship, Lamont-Doughy Geological Observatory
1985-87	Peck Foundation Merit Award, Beans University

Peer Reviewed Scientific Publications

(Insert here a list of 33 peer-reviewed publications (16 as first author), 15 technical reports/data packages, and abstracts since 1996.)

Case Study Number 4: Harriet Dean Stanchion

As a postdoc, Harriet Dean Stanchion had become the computer guru for the entire geophysics group at Los Aimless National Laboratory. Best of all, she found that she liked it! She enjoyed the intellectual challenges of making programs run and keeping complex computer systems operating. She liked the fact that, while the overall job stayed the same, there were always new projects and challenges to face. She also liked the fact that she worked with people, solving their problems and helping them get their work done.

But as the end of Harriet's postdoc neared, she realized that her future was not as certain as she had hoped. While she had become an indispensable member of the geophysics group at the Lab, she had only published a few papers in her 3 years as a postdoc. This greatly diminished her chances of being hired as a permanent member of the scientific staff.

Harriet was sure that she wanted to work with computers, but she was uncertain what sort of jobs were out there. She was sure that she did not want a job in which she would be programming all day: she preferred a combination of tasks and challenges that included some programming. Most importantly, she wanted a job in which she would work with people.

Harriet began her exploration by talking to friends and friends of friends in the information technology sector. She learned that there were a number of jobs open, but when she perused the want ads or job descriptions she found that those openings called for one activity only-either programming or systems management, but not both. In addition, the jobs she saw advertised tended to be lower-level positions. While they paid quite well, Harriet was fairly sure that they would not offer the level of challenge that she sought.

At the same time Harriet was exploring the outside world, she was busy trying to make the bureaucracy at her current job realize that she was indispensable. She discussed her future with her supervisor and with the head of the geophysics group. Both wholeheartedly agreed that Harriet was an asset and that the group would suffer if she left. Harriet urged them to discuss the problem with upper management. She wrote a memo describing all the projects and improvements she had carried out while acting as an informal systems administrator. It was an impressive list. She forwarded the memo to her supervisors, who then used it in their arguments with management.

Finally, the powers-that-be responded. They wanted to see Harriet's resume. This resume was to be examined by supervisors in the computer support department, not the scientific staff. Harriet was panic-stricken. What should she do? Harriet took her CV and pared it down to a single page and included some information about her past work experience outside of geophysics. She then took this resume to a friend who worked in a computer hardware company and who had hired people in the past.

Harriet's friend liked what he saw, but found the style of the resume wanting. The paragraph style in Times 10-point font was hard to read and the resume looked cramped. He suggested several changes including dropping the personal interest section (didn't look professional), strengthening the objective statement to better fit the potential opening (or dropping it altogether), and changing the font and layout to make it more readable. Harriet's first resume is on the next page.

Harriet set about revising this resume. In places, she beefed up the job descriptions and used numbers to demonstrate her points. Harriet also emphasized her project management experience because she knew that the job that might open up for her would require some experience handling budgets, purchasing equipment, and overseeing contracts. She changed the layout to improve readability and squeezed the whole thing onto one page. The new resume (on the following page) looked much better.

Serendipity Strikes

As Harriet explored the field of computer management and systems administration, she learned that the jobs that involved multiple tasks, the sort of jobs she was after, were more likely to be found in a smaller company rather than the big firms whose ads she saw in the newspaper. Small start-up companies needed versatile, team-oriented people to run their computers. And, for those people who were lucky enough to hook up with a small company that grew, the monetary rewards were enormous.

Harriet began to network in earnest, calling friends, researching companies on the Internet, and trying to find out which small companies might be in the market for a systems administrator. She got several names, made some contacts, and was all ready to meet some people.

Then a funny thing happened.

Harriet's Mom was a realtor with a major real estate company. When Harriet returned home for Thanksgiving her mom told her of an interesting development at her work. It seems that several national real estate companies, including hers, were using the Internet to market property directly to consumers. Her company was just getting the project organized but, as they had very little experience in network computing, they were proceeding slowly.

Harriet met with the regional VP in the company and heard about the company's goals. The project was immense, involving multimedia presentations, complex client/server environments, and state-of-the-art Web technology. If there was a job, it would be as project manager, supervising a staff of 6 and handling a budget of $1,200,000 in the first year. The VP

Harriet Dean Stanchion
Los Aimless National Laboratory
L-666 700 West Avenue • Los Aimless, New Mexico 99321
(505) 555-7865 • (505) 555-4567 Fax • stanton@s145.ep.lanl.gov • http://www.ssshake.com

OBJECTIVE: To be a key contributor in a challenging project with opportunities in problem solving, communication of ideas, management of information and computer resources.

EDUCATION:

1997 Ph.D., Geophysics, **Mighty University**, Bigville, California
Dissertation: ***Tectonics of Western North America: A Seismic Perspective***

1990 B.S., with highest distinction, Geophysics, **Boston University**, Boston, Massachusetts
w/course work in analog & digital design, filter design, DSP

SKILLS:

Machines	Sun SPARCstation 2,10,20, Sun 3's, Sun 4's, DEC VAX, 5000's, Macs 030, 040, PPCs
O/S	Solaris 2.x, SunOS 4.x, Ultrix 4.2, BSD 4.3, IBM AIX 3.2, MacOS 7.x, 6.x
Equipment	REFTEK 16-bit dataloggers, Kinemetrics /Guralp seismometers, Trimble GPS receivers
Software	PROMAX, PASSCAL(IRIS), SAC/MAP, Adobe Photoshop & Illustrator, MacHTTP & NCSA WWW servers, Mosaic, Netscape, Forward & Inverse seismic modeling codes
Languages	FORTRAN, BASIC, assembly

EXPERIENCE

1/97 - present **Research Scientist, Technical Illustrator, Sysadmin**
Los Aimless National Laboratory, Los Aimless, NM

Project Manager - Northern Sierra Nevada Seismic Study, 14 month project duration. Responsibilities included: design and implementation of experiment and data collection efforts, including merge of five types of seismic data totaling 1 GB, data analysis, coordination of the 15 collaborating Universities, private, local, state, and Federal agencies, task a 75+ person field crew, $600K budget. Collaboration in the LANL/Columbian South American seismic study involved data collection, analysis, and data archiving (30 GB seismic databases). Presented scientific results frequently in both oral and written form including publication in both peer-reviewed journals and electronic media (WWW and E-documents). Hardware , Software, and Technical illustration consultant(Sysadmin) for network of SUN/DEC/IBM workstations and Macintoshes. Volunteer for local school (geoscience lectures, computer advice).

9/90 - 12/96 **PhD candidate - Research Assistant, Sysadmin (Geophysics Dept.)**
Mighty University, Bigville, CA

Provided computer purchase recommendations, installation, and computer support for networked UNIX workstations and software tools for 100+ faculty and students in the Geophysics Department. Included cross-platform (SunOS, DEC, Mac) software tool installation and support for graphics and scientific software. Networked sharing and load balancing of cpu and disk resources. Designed and implemented the disaster recovery plan for department computer mainframe and workstations. Plan used successfully to recover from Loma Prieta earthquake. Provided assistance in technical illustration and Internet information services.

8/86-4/90 **Tutor math, physics, and geoscience courses**
Boston University

Developed teaching aids and course material for a variety of geoscience, physics, and electrical engineering classes. Planned field trips to augment geoscience courses. Worked with students with differing levels of background knowledge and various skill levels.

6/89-8/89 **Electronic Design Technician**
Q.U.I.T. Inc., Sticks, Virginia.

Developed and implemented an innovative procedure for testing computer control systems and provided feedback to the manufacturing division.

6/85-8/88 **Electronic Design Technician,**
Novel Applications Inc. Burbs, Va.

Responsible for the computer control system hardware (MC6800) and software(assembly) for a DOD nuclear pulse thermal test facility. Provided system documentation and training for users. Design considered scientific and safety (control of high voltage energy storage system) constraints.

STRENGTHS: Effective communicator, Problem definition/solving, Team building/collaborations
INTERESTS: Quiet mountain hikes, bicycling, education
REFERENCES: Available upon request

Harriet Dean Stanchion
Los Aimless National Laboratory
L-666 700 West Avenue • Los Aimless, New Mexico 99321
(505) 555-7865 Tel. • (505) 555-4567 Fax • stanton@s145.ep.lanl.gov • http://www.ssshake.com

EDUCATION

Mighty University Ph.D. in Geophysics **1997**

Boston University B.S. in Geophysics–*Graduated with highest distinction* **1990**
Extensive coursework in Electrical Engineering

TECHNICAL EXPERIENCE

Research Scientist, Project Manager and Systems Administrator **1997 - Present**
Institute of Geoplanetary Physics, Los Aimless National Laboratory, Los Aimless, New Mexico
Northern Sierra Nevada Seismic Study: Managed 14-month scientific study across 3 states. Designed and implemented experiment, data collection and analysis. Coordinated efforts of collaborating universities and private foundations. Directed 75+ person field crew and oversaw a $600K budget. Co-authored 3 papers for publication and presented 4 papers at national and international conferences.
LANL/Colombian South American Seismic Study: Organized data collection, analysis, and data archiving. Supervised and assisted in the installation and maintenance of 15 seismometers across Bolivia. Prepared and maintained HTML documents describing project and data.
System Administrator: Hardware, software, and technical illustration consultant for network of 16 SUN/DEC/IBM workstations and 20 Macintoshes. Provided educational lectures and computer network advice to local schools.

Research Assistant and Systems Administrator **1990 - 1997**
Department of Geophysics, Mighty University, Bigville, California
Provided computer purchase recommendations, installation, and computer support for networked UNIX workstations and software tools for 100+ faculty and students in Geophysics Department. Installed and supported cross-platform (SunOS, DEC, Mac) software tools for graphics and scientific applications. Coordinated network sharing and load balancing of CPU and disk resources. Designed and implemented a natural disaster recovery plan for department computer mainframe and workstations that recovered 100% of computer capabilities after Loma Prieta earthquake. Assisted in technical illustration and Internet information services.

Electronics Design Technician **1985 - 1988**
Novel Applications Incorporated, Burbs, Virginia
Assembled and tested computer control system hardware (MC6800) and software for DOD nuclear pulse thermal test facility. Wrote and produced system documentation and training for users. Implemented scientific and safety control of high voltage energy storage system constraints in designs.

COMPUTER SKILLS

Machines	Sun SPARCstation 2,10,20, Sun 3s, Sun 4s, DEC VAX, 5000s, Macs 030, 040, PPCs
O/S	Solaris 2.x, SunOS 4.x, Ultrix 4.2, BSD 4.3, IBM AIX 3.2, MacOS 7.x, 6.x
Software	PROMAX, PASSCAL (IRIS), SAC/MAP, Adobe Photoshop & Illustrator, MacHTTP & NCSA WWW servers, Mosaic, Netscape, Forward & Inverse seismic modeling codes
Languages	FORTRAN, BASIC, assembly, functional in C++ , PERL, and Pascal.

Harriet Dean Stanchion

342 Oppie Lane • Los Aimless, New Mexico 99321
(505) 555-7865 • (505) 555-4567 Fax • stanton@s145.ep.lanl.gov • http://www.ssshake.com

SUMMARY: Versatile, creative, entrepreneurial information systems developer with excellent organizational, analytical and computer skills, 3 years of project management experience, and familiarity with commercial real estate practices.

SYSTEMS MANAGEMENT EXPERIENCE:

Systems Administrator **1997 - Present**
Institute of Geoplanetary Physics, Los Aimless National Laboratory, Los Aimless, New Mexico

- Hardware, software, and application consultant for network of 16 SUN/DEC/IBM workstations and 20 Macintoshes.
- Maintain, repair, and upgrade system hardware and software. Develop and modify scientific applications.

Research Assistant and Systems Administrator **1990 - 1997**
Department of Geophysics, Mighty University, Bigville, California

- Provided computer purchase recommendations, installation, and computer support for networked UNIX workstations and software tools for 100+ faculty and students in the Geophysics Department.
- Coordinated network sharing and load balancing of CPU and disk resources.
- Designed and implemented natural disaster recovery plan for department computer mainframe and workstations that recovered 100% of computer capabilities after Loma Prieta earthquake.
- Assisted in technical illustration and Internet information services.

PROJECT MANAGEMENT EXPERIENCE:

Research Scientist, Project Manager **1997 - Present**
Institute of Geoplanetary Physics, Los Aimless National Laboratory, Los Aimless, New Mexico
Northern Sierra Nevada Seismic Study:

- Managed a 14-month scientific study across 3 states. Designed and implemented experiment, data collection, and analysis.
- Prepared and presented permitting cases for local, state, and federal agencies.
- Directed 75+ person field crew and oversaw a $600K budget.

LANL/Colombian South American Seismic Study:

- Organized data collection, analysis, and data archiving.
- Supervised and assisted in the installation and maintenance of 15 seismometers across Bolivia.
- Prepared and maintained HTML documents describing project and data.

COMPUTER SKILLS:

Machines	Sun SPARCstation 2,10,20, Sun 3s, Sun 4s, DEC VAX, 5000s, Macs 030, 040, PPCs
O/S	Solaris 2.x, SunOS 4.x, Ultrix 4.2, BSD 4.3, IBM AIX 3.2, MacOS 7.x, 6.x
Software	PROMAX, PASSCAL(IRIS), SAC/MAP, Adobe Photoshop & Illustrator, MacHTTP & NCSA WWW servers, Mosaic, Netscape, Forward & Inverse seismic modeling codes
Languages	FORTRAN, BASIC, assembly, C+ +, PERL, Pascal.

EDUCATION:

Mighty University	Ph.D. in Geophysics	**1997**
Boston University	B.S. in Geophysics–*Graduated with highest distinction* Extensive coursework in Electrical Engineering	**1990**

explained that they were looking for a versatile, adaptable team-player, a "visionary with excellent computer skills." Harriet described her background and experience and was asked to fax the VP her resume the next morning.

Harriet spent the rest of the afternoon revising her resume for the possible job opening with the real estate company. In her discussion with the VP, Harriet heard several specific skills that they needed. First and foremost was extensive experience with networked systems, system administration, and programming ability in C++ and PERL. The other concern she detected from the VP was that the person involved have project management experience and some familiarity with the Real Estate industry. Harriet had good project management experience and knowledge of the real estate business she osmosed from her mother. Harriet revised her resume accordingly (see page 130), removing some of the extraneous science, emphasizing the project management experience, and explaining that she had a good working understanding of the real estate business (though she didn't say from where).

Case Study Number 5: Richard Fineman

Richard Fineman has, by all accounts, an outstanding record of scientific achievement. As an undergraduate student on the west coast, Richard carried out two undergraduate research projects with an outside advisor and graduated with highest honors in physics. He went to a famous Ivy League university to earn a Ph.D. with a mighty famous astrophysicist and, having done well there, landed a plum postdoc at a national laboratory. His research was innovative and he had a strong publication record and outstanding teaching ability. Even in this current grim job market in astronomy, Richard should have a good shot at a research faculty job.

If that were his goal in life.

In fact, Richard was about to step away from the world of research science, possibly forever. His decision does not stem from frustration, unhappiness, or dissatisfaction with his life as a scientist. He was simply ready to try something new. Richard was about to embark on a career in investment finance.

Was it a lust for money that drove him to this career change? Was he intoxicated with greed? Driven by a hunger for a Gucci-shoed, BMW-driving, single-malt-scotch-sipping lifestyle? No. Richard came to the conclusion that he liked quantitative analysis of financial markets the old fashioned way: he learned it.

As a graduate student, Richard had two friends who went off to the world of business, one as a management consultant, the other as an investment analyst. Contrary to the predictions of his advisor, the mighty famous astrophysicist, Richard's two friends did not become depressed, money-obsessed yuppies, but remained very much themselves. In fact, both of them loved their jobs.

When Richard graduated with his Ph.D. and began his postdoc, the idea of doing something different with his life remained. However, unlike his two friends, Richard both enjoyed and wanted to continue using his expertise in computational modeling and analysis. He considered the possibility of a programming job but didn't feel comfortable being a cog in a large machine: he enjoyed working independently.

Serendipity and the prepared mind intersected the day Richard decided to audit a course at a local university on new computational methods for pattern recognition. Richard's original intention was to apply the material from the course to his current research modeling X-ray emissions from active galactic nuclei. However, when he arrived at the class, he learned that the instructor and most of the people there were interested in financial modeling. After 3 weeks in the class, Richard was interested, too.

Richard did very well in the class and carried out two extra projects with the instructor. After the class ended, the instructor asked Richard if he would be interested in trying a bit of outside consulting. Richard leapt at the chance to apply his computational tools to a new set of problems. Earning $60.00 an hour for a week didn't hurt, either.

One consulting job rolled into another and another. Richard realized that he enjoyed what he was doing as much as he enjoyed doing science, and he had greater flexibility and could work at home. He made arrangements with his postdoc supervisor to go part-time, and spent the remainder consulting. In the course of 1 year he and his wife had their first child, bought a house, and began to settle in.

After consulting part-time for a year, Richard desired a more permanent arrangement with his clients. He prepared a formal proposal to the president of Tempest Research and Trading, his main client. He made the case that the company needed him full-time, not just to carry out modeling, but to assist all the traders with applying quantitative analyses to their work.

He included his resume (opposite page) with the packet of materials and received a call the next day. The company was interested; they would forward the proposal to Chicago, where the business office would approve the position. There was only one problem. The resume he included looked "weird" to the business types. Could he send a modified version?

Richard showed his resume to his wife and to two friends, and to one of his friends who worked for the company. Their suggestions included everything from changing the format to including more information. Richard followed their suggestions fully, correcting some typos, and produced a new resume.

This resume looked much better. It looked professional and followed the format and style that the business people expected. It clearly demonstrated his technical skills and the communication skills needed to interact with the rest of the company. It described more fully his past achievements in language that business people could understand.

Richard ended up getting the job, and, after 2 years, struck out on his own, founding a company that performs quantitative analysis of futures markets.

Richard D. Fineman

3455 Tamirand Ct.
Pleasureville, California 95678

SUMMARY:

- Artificial Neural Networks: Theory and applications to pattern recognition and time series forecasting. Potential applications to finance and portfolio management
- Finance: Pricing Models for Derivative Securities
- Scientific Programming: Differential Equations, Hydrodynamics, Neural Network Training.
- Theoretical Cosmology and Astrophysics, Advanced Mathematics.
- Oral and written technical presentations. University level teaching.

EXPERIENCE:

Consultant, Tempest Research and Trading, Tempest, CA 6/97-present

- Develop algorithms and write programs in C++ for pricing of derivative securities.
- Black-Scholes, binomial tree, Heath-Jarrow-Morton methodology, Monte Carlo methods.

Postdoctoral Research Fellow, Los Aimless National Laboratory 9/96-present

- Pure and applied research in astrophysics, astronomy and neural networks.
- Neural network applications including pattern recognition and time series forecasting
- Analytical and numerical methods. Scientific programming. Unix.
- Organize and run weekly technical seminars at LLNL
- Led neural network research team in LLNL Summer Undergraduate Research Institute.

Postdoctoral Fellow, Standish Astrophysical Observatory, Our Fair City, MA 6/96-9/96

Graduate Research Fellow, Hallowed-Standish Center for Astrophysics 9/91-6/96

Research Assistant, Space Sciences Laboratory, UC Tempest 3/90-9/91

Research Assistant, NASA/Fires Research Center, Muffler Field, CA 9/86-3/90

Education:

- Hallowed University, Ph.D. Physics, 6/96
- Hallowed University, M.A. Physics, 12/92
- University of California, Tempest, B.A. Physics, 6/91

HONORS:

- Graduated with highest Honors (UC Tempest 06/91)
- Member Phi Beta Kappa
- Regents' Scholar (UC Tempest)
- Alumni Scholar (UC Tempest)

Richard D. Fineman

3455 Tamirand Court
Pleasureville, California 95678
Tel: (530) 555-9001

SUMMARY

- Extensive knowledge of Neural Network theory and applications to pattern recognition, time series forecasting, and pricing models for derivative securities
- 8 years programming experience using C and C++
- Working knowledge of derivative securities markets
- Excellent written and oral communication skills

EXPERIENCE

CONSULTANT, TEMPEST RESEARCH AND TRADING **1997-present**
Tempest, California

- Developed algorithms and wrote software programs in C++ for pricing of derivative securities using Black-Scholes, binomial tree, Heath-Jarrow-Morton methodology, and Monte Carlo methods.
- Assisted traders with quantitative assessment of pricing strategies.
- Coordinated distribution of pricing information to traders in Chicago and Vancouver.
- Advised senior management regarding novel applications of numerical analysis.

RESEARCH FELLOW, LOS AIMLESS NATIONAL LABORATORY **1996-present**
Los Aimless, New Mexico

- Developed novel computer applications of neural networks to pattern recognition and time series analysis of astrophysical data.
- Managed 3 UNIX workstations.
- Led 12-week neural network research team (LANL Summer Undergraduate Research Institute) for 6 undergraduate interns.
- Published 11 peer-reviewed papers and reports and presented 10 papers at national and international meetings.

RESEARCH ASSISTANT, HALLOWED-STANDISH CENTER FOR ASTROPHYSICS **1991-1996**
Our Fair City, Massachusetts

- Developed theoretical and computational methods for constraining models of cosmic X-ray and infra-red background emission and successfully modeled the observed X-ray emission from active galactic nuclei.
- Collaborated with scientists from Hallowed University and the Swedish Astrophysical Station (Downsalla, Sweden).
- Taught 1 graduate course and 2 undergraduate courses in Physics.

EDUCATION

Ph.D and M. A. in Physics, Hallowed University, Our Fair City, Massachusetts **1996**
B.A. in Physics, University of California at Tempest, Tempest, California **1991**

HONORS AND AWARDS

Graduated with highest honors (Summa Cum Laude), U.C. Tempest (1991)
Member Phi Beta Kappa (Elected 1991)
Regents' Scholar, U. C. Tempest (1990-91)

Case Study Number 6: Karen Smote

By the time Karen Smote finished her Ph.D. she was so sick of her thesis and graduate school that she was ready to depart the world of science for good. While her unpredictably aloof and bullying advisor certainly hadn't made her life any easier, most of what Karen didn't like about graduate school was, well, the research. She enjoyed the computer work and programming she carried out as part of her research but the research topic left her cold.

However, Karen's partner, Susan, had a great time in graduate school and was ready and eager to pursue a research career. The two graduated within 6 months of each other and Susan landed a good postdoc in New York City. Only 6 weeks after finishing her thesis, Karen found herself on the east coast in a new city with no job.

Karen had received plenty of advice from friends about the importance of self-assessment and finding her life goals and work passions. That was all well and good, but such ruminations took time, and Karen needed a job. She knew little about the world of work outside of academia and did not have the luxury of time to explore her options leisurely. Despite the admonitions of career books and counselors, Karen looked for a job the old fashioned way: she sent out scores of resumes and cover letters in response to job advertisements in the newspaper.

Karen targeted entry-level computer administrator or UNIX system administrator jobs. She had 5 years experience working with UNIX systems and Macintosh computers in grad school, and she enjoyed doing it. There were opportunities for employment in this line of work in all sorts of companies, large and small, profit and non-profit. Karen wanted to find a job that would allow her to build her computer skills to enable her to get another job quickly, as Susan's postdoc was for only 1 year and they were likely to move.

Karen put together her resume, accenting her technical experience, her computer skills, and her ability to organize and manage projects independently. This combination of skills and experience might look unusual to those hiring entry-level computer managers, but it would show a record of accomplishment. Karen applied to 24 advertised positions with this resume.

Karen waited. After 2 weeks of not hearing anything she began to worry. Was the job market really this bad, or was it her? Finally, one manager called her. "Did you mean to send your resume to me, or was this an accident? Your background is impressive, but this job is entry-level and not suited for someone with an advanced degree."

Karen explained that, while she did have a Ph.D., she had found computer work the most enjoyable side of her research. She explained that she was in the process of switching careers and was seeking an entry-level position in which she could gain experience and advancement. The manager on the other end of the line seemed more interested after hearing this and she eventually got an interview. But she didn't get the job.

Karen Smote

200 Jude Avenue, #233
Englewood Cliffs, NJ 07654
Tel: (212) 555-3433
E-MAIL: ksmote@gaea.uu.edu

EDUCATION

Struggling University, Pullback, WA, Ph.D. in Chemistry 9/94-9/99

Thomas Jefferson University, Washington, D.C., BS in Chemistry (Physics minor) 8/90-5/94

COMPUTING EXPERIENCE

Environmental Engineer, Dames and Broads Consulting, Pullback, WA 12/98-8/99

- Used MS-DOS-based computers and software such as, Word, Excel, and CC:Mail for Windows
- Programmed in dBase
- Used ChemInfo and USGS WATER-CRUD software to create and customize water quality databases for hazards modeling
- Wrote documents detailing methodology used in database development

Lab Manager, Struggling University, Department of Chemistry, Pullback, WA 7/98-8/99

- Maintained and operated an ICP-Mass Spectrometer
- Maintained a dedicated DEC PDP11/23+ (running RSX-11M operating system) used to calibrate and run ICP-Mass Spectrometer
- Instructed lab users in computer and spectrometer use

Research/Teaching Assistant, Struggling University, Department of Chemistry, Pullback, WA 10/94-8/99

- Used DOS-based computers for running various geochemical programs
- 5 years experience with UNIX-based, networked computer system, including e-mail, FTP, telnet, and Gopher on the Internet
- Acquired extensive experience with Macintosh computers and software, including CricketGraph, DeltaGraph, EndNote, Eudora, Freehand, Ilustrator, MacDraw, Mosaic, NetScape, StatView, and Word
- Use of Montage hardware and POM and ImageQ software to prepare photographic slides for presentation and national conferences
- Taught undergraduate and graduate laboratory sections for various introductory and upper-division classes
- Prepared and submitted research results for publication in peer-reviewed journals

ADDITIONAL INFORMATION

- Programming ability in Basic, Turbo Pascal, and FORTRAN
- Experience with NetScape and Mosaic (World Wide Web browsers)
- Reading proficiency in Spanish
- Volunteer for the American Red Cross
- Interests include: rock climbing and backpacking

Karen Tries a Skills Resume

After a month in which she received only two responses to 24 solicitations, Karen decided that her resume was clearly not working. She picked up a resume-writing book and re-engineered her chronological resume into a functional resume. This new resume itemized her particular computer skills and experience and presented her work history at the bottom. While it still demonstrated her record of accomplishment, her skills were more prominent. She didn't completely remove evidence of her Ph.D., but by placing it at the bottom, it occupied a less prominent position. Karen applied for 25 positions with this new resume, and almost immediately detected a difference. By the end of 2 weeks she had eight positive responses and several interviews.

As Karen interviewed for some of the jobs, she learned how her two resumes had been treated differently by the various companies to which she had applied. Her first resume, while it contained information about her computer skills, emphasized parts of her work history that were totally outside the requirements and background of the positions to which she applied. Rather than stimulating interest, employers seemed to reject her resume immediately as being too unusual. This was even more true if her resume had gone to Human Resource departments rather than to the managers advertising the job. Human Resource departments, Karen learned, were often somewhat uninformed about the technical nature of the jobs they were trying to fill, and simply looked for those candidates whose skills matched those listed on the job description. The two people who had contacted Karen after receiving her first resume did so simply because they were intrigued, and perhaps even a little worried, that her resume had been sent to them by mistake.

Karen's second resume (see following page) was treated much differently. By emphasizing her skills over her unusual background, it seemed to receive more scrutiny. Her advanced degree, while obvious on her resume, assumed a far smaller role in the qualifications she presented. Karen had chosen to switch from a chronological to a skills resume because she read that skills resumes were more effective for people making radical career shifts, or for those who didn't have a continuous work history. According to the experts, skills resumes made an unusual background look more compatible. Karen's experience was a spectacular confirmation of this.

Karen Encounters the World of Work

Karen interviewed with 10 employers. While the needs of each company were similar, the companies themselves were wildly different. Two were finance companies on Wall Street, three were manufacturing companies, three were service companies, including a "Renter's Digest" publishing firm, and two were non-profits. The work atmosphere she encountered in each place was distinctive. The non-profits had a relaxed dress code and a frantic, low-budget atmosphere that reminded Karen of academia. The finance companies had a strict dress code: suits and dresses. They were more formal, but also more opulent. They also paid better! Karen found the non-profit work environment appealing. She was turned off by the financial world, especially the dress code.

Karen Smote

200 Jude Avenue, #233
Englewood Cliffs, NJ 07654
Tel: (212) 555-3433
E-mail: ksmote@gaea.uu.edu

MACINTOSH EXPERIENCE

- Application software: Cricketgraph, Canvas, ClarisWorks, Deltagraph, Endnote, Freehand, ImageQ, MacDraw, MS Word, MS Excel, Professional Output, Manager, StatView, Stuffit, SuperPaint
- Internet-access software: ConfigPPP, eWorld, Eudora, Fetch, MacIP, MacTCP, Mosaic, NCSA Telnet, NetScape, TurboGopher, ZTerm
- Configured modem for use with shell and PPP Internet accounts

DOS AND WINDOWS EXPERIENCE

- Application software: MapInfo, MS Excel, MS Word, WordPerfect, WordStar
- Internet-access software: CC:Mail, Mosaic
- Created customized databases for use with proprietary, Windows-based software on DOS-based LAN

OTHER COMPUTER SKILLS

- Programmed and debugged code in dBase, FORTRAN, Turbo Pascal, and Basic
- 5 years experience with UNIX including vi, mail, telnet, ftp, shell scripts, awk, LaTek
- Experience with HTML
- Maintained and repaired DEC PDP11/23+ computer

COMMUNICATION SKILLS

- Presented research findings orally and in poster format at 6 national conferences
- Documented database development methodology
- Published 2 papers and 12 abstracts in peer-reviewed journals
- Taught complex scientific principles to university students through lectures and demonstrations

ORGANIZATION SKILLS

- Managed analytical laboratory—scheduled lab usage, trained lab users, ordered supplies, compiled quarterly billing, kept lab in compliance with Health and Safety codes
- Organized Fall and Winter Journal Club speaker series at Impressive University
- Hired and coordinated scheduling of lifeguarding staff at 40 swimming pools

EMPLOYMENT

1998-1999 Dames and Broads Consulting, Inc., Pullback, Washington–Part-time Assistant
1994-1999 Struggling University, Pullback, Washington–Lab manager; Research/Teaching Assistant
1990-1992 Thomas Jefferson University, Washington, D.C.–Reserve Desk (Gelding Library)
1990-1994 Hippocampus Pool Service, Falls Church, Virginia–Staffing Supervisor

EDUCATION

Struggling University, Pullback, Washington, Ph.D. in Chemistry **1994 - present**

Thomas Jefferson University, Washington, D.C., B.S. in Chemistry (Physics minor) **1990 - 1994**

But as her interviews progressed (for some employers she ended up having three interviews), she found that the world of finance offered other advantages. One company she interviewed with had a 3-day, 12-hour work week. While this would mean three long days, it also allowed Karen 4 days off a week. In addition, the financial company had in-house computer training for their employees and reimbursed their employees for courses they took outside the firm. For Karen, who wanted to increase her skills and marketability, this combination of free time and subsidized learning was too good to miss. When they offered her the job, she took it.

Conclusion

The resume is not the only important part of your job search, but it is a very important device for finding a job. Where you send your resume is as important as how its written. Each of the people profiled in this chapter landed a job because they presented themselves well AND they knew who to present themselves to.

Summary

- Serendipity is an important part of a job search. Be prepared to recognize opportunities and act fast by adapting your job search materials to each opening.
- While the content of your resume is the most important element, a clear and readable layout always helps!
- Review the resumes of friends or a book of resumes from a job bank to get ideas about layout and content.

Listen to this excellent cover letter: "I, Grotok, Overlord of Mulingek, demand a position of power at your puny company..." We've GOT to hire this guy!
AJL

Cover Letters

Going from Huh? to Wow!

11

If you think resumes get jobs, think again. Resumes are only as good as the letter accompanying them. Preoccupied with writing good resumes, most job seekers fail to properly introduce their resumes to potential employers. Be forewarned: neglect your cover letter and you may quickly kill your resume!

Ronald Krannich and Caryl Rae Krannich
Dynamite Cover Letters

Cover letter. I hate the term "cover letter." The word implies that the document you produce to accompany your resume and other job materials, your opening salvo, the first volley of information that you present to a prospective employer, is simply gift wrapping for your resume. Sadly, most people treat it that way. While they may spend hours sprucing up their resumes, their cover letters may receive only cursory attention. And when the prospective employer opens the mail and pulls out the resume and cover letter, the first reaction is: ho hum.

A cover letter should be treated with the care and precision with which you would treat the first three paragraphs of a huge proposal to the NIH or NSF. The cover letter is your proposal, your sales pitch, for why they should hire you instead of anyone else. It should address the core of your qualifications and should guide the reader into your resume. It should answer their questions and pique their curiosity about you. And above all, it should show your flawless command of the written word and your knowledge of proper business letter etiquette. If it does not it amounts to little more than a fish-wrapping implement.

A good cover letter is like bonus points—pulling your grade up from a B+ to an A.

While a good cover letter may not dramatically change your chances, a lame cover letter may really hurt you. Misspellings, sloppy grammar, and inattention to business letter style are all caution flags that an alert reader will notice. Such errors suggests carelessness or, worse, incompetence! Because your cover letter is the first example of written English that a prospective employer may read, it is critical that it be as perfect as possible.

Content is also important. Writing the perfect cover letter is hard because it requires anticipating the main concerns of the person who reads it. If you have done some research on the company, had an informational interview or two, talked to your network, and spoken with the hiring manager you will probably know all the inside issues surrounding the job opening. If you have not carried out any research and are shooting in the dark, your letter may be off target.

Structure of the Cover Letter

A professional-looking letter starts with good paper. Expensive cotton bond is not necessary, but something a notch or two above stock Xerox paper is preferable. You can choose any color paper you want, so long as it is white. Don't bother with textured paper; not only will it look weird but your laser printer might not be able to fix the toner onto it securely. The last thing you want to send a prospective employer is a letter that looks like a Rorschach test!

Should you use departmental stationary? It depends on your status and the policies of the department. If you are presently an instructor or a member of the research "staff" (I would think that postdocs fall into that category) then it is appropriate to use official letterhead. If you have any question about this, ask the departmental administrator or the department chair. If, however, you are no longer supported by the department or institution, or the return address is your home, then using the department letterhead might seem fishy.

One-page cover letters are a RULE, unless you have some specific reasons to make them longer (for example, if the job description or advertisement ASKS for answers to questions or more information). In fact, three paragraphs should be sufficient. If your letter is any longer, you'd better have a good reason for it. Here is a suggested general structure:

Paragraph 1:

- introduce yourself to the reader
- explain why you are writing (either for a specific opening or a potential opening)
- explain how you learned about the position
- explain why you'd be perfect for the job

Paragraph 2:

- show how your qualifications fit the job
- demonstrate your suitability by citing examples
- expand on one or more items from your resume that highlight your key qualifications

Paragraph 3:

- state what the next step is (for example, you will call in a week to check up)
- thank them

In the first paragraph, you should say why you are sending your resume. Are you applying for an advertised position or just a potential opening? If it is a specific opening, where did you learn about it? You would be surprised how many people fail to mention the specific job for which they are applying. For big companies that are advertising many positions, your lack of specificity may land your resume in the recycle bin. Even for small operations, it is important to explain how you heard about the job.

Most mediocre cover letters are not specific. They cite items in the resume but they fail to make the connection to the job that is being advertised. Yeah, sure it's a great thing that you have worked for a summer doing data reduction but what if they are most concerned about project management experience? You're hosed! You have to show them how your background and experience fit the job they are advertising. At the bare minimum, you should have the job advertisement in front of you as you are writing. But you will be further ahead if you've actually done some research on the company or, better yet, have talked to the people who are advertising the opening.

The best way to prove that you are a good fit is to cite examples in your past work history in which you tackled similar job duties or occupied a similar position. Do NOT assume that they will pick these details out of your resume, especially since the average employer spends only 20 seconds scanning through it. For example, if you were applying to work in an aeronautical engineering company doing product development you want to note any specific experience involving the development of a device or experiment, rather than just citing your years of experimental work for your Ph.D. Citing specific examples and quantifying them where possible is the best way of convincing a stranger that you've got what it takes to get the job done.

The final paragraph of the cover letter needs to be a proposal for action. You need to state what you would like the next step to be. For example, if you are applying for a specific job opening with a deadline, you may want to state that you will call before then to confirm that they have received your job materials. Alternatively, you may state that the employer is free to call you with any questions.

Finally, think like a marketing executive. Understand your story and your message. Think carefully about how to communicate your message, and do it in language that your audience will understand. Be clear, compelling, and concise.

Issues for the Scientist Applying for a Non-science Job

Scientists applying for non-science jobs may want to address additional issues. One of the most obvious questions that might come up in the minds of the people reading your resume is:

> *Why is a scientist applying for this job?*

Some related points of concern may be:

Is this person over-qualified?
Why is the experience as a scientist valuable in this position?

These questions, if not addressed clearly, may disqualify a scientist job applicant right off the bat. It is important to EXPLAIN why you as a scientist are applying for a non-science job. Reassure the reader that you are not clueless, or worse, desperate, and that you have a genuine desire to work for the company or oranization. An employer reading your resume may come to the education section, see your Ph.D. or Masters degree, and say, "Huh?" After reading your cover letter, the employer should say: "Wow!"

Cover Letter Writing Tips

Here are some final bits of advice to use when you write your cover letter:

1. ***Use their words.*** Match their job description to your background. If they say that the job requires "project management experience," use those words. Depending on the length of the job description, you may be able to do a point-by-point response to each item listed in their job description. Don't copy all of the key words of the job ad verbatim, just make certain you have made the connection between their needs and your abilities.

2. ***Write to a person, not to a human resource department.*** In some cases, it is simply unavoidable to do an end-run around the HR department. But if possible, direct your job materials to a person, ideally the person making the hiring decision. You might also consider sending a copy of your cover letter and resume to the person or people with whom you have had prior contact in the organization to alert them that you have made a formal application for employment. This gives them the opportunity to walk down the hall and put in a good word for you if they are so inclined.

3. ***Be concise.*** Your life story may be terribly interesting but nobody is likely to turn the page. Make a one-page letter the rule unless specifics dictate otherwise. This shouldn't be too hard for you. After all, you've spent years cramming research results and conclusions into abstracts no bigger than a fig leaf. So you know you can cover a lot with a fig leaf.

4. ***Stress the positive.*** Do not, under any circumstances, lay out a sob story. Don't tell them that your postdoc is running out, or that you'll be kicked out of the country in 6 months when your F-1 visa expires. Those ugly details will need to be discussed if and only if you get an interview and an offer. Candor is good... to a point.

5. ***Avoid cover letter clichés.*** While it is important to thank people for spending the time to read your letter, you would be wise to develop a novel way of saying this besides "thank you for your consideration." Yeck! What a boring thing to say! How wooden! Say it in your own words or don't bother. Also, try to avoid "enclosed please find" and "attached is my resume." These are clichés: boring over-repeated drivel that is the sign of a lackluster writer! Show them your uniqueness: be your own dog!

For those of you for whom English is not your first language, this jihad against clichés may seem a bit unfair. After all, these stock phrases are new to you, and may fall slightly more trippingly off-the-tongue than most English. Try this instead: give your letter to a friend who is a native English speaker and ask him or her to circle any clichés they find. Then discuss with them how you might express yourself differently. Non-native speakers should pay particular attention to their cover letters because it is, among all the other things mentioned above, a sample of your best writing.

Cover Letters: A Before-and-After Example

There are numerous examples of cover letters available to you—in books, at career planning and placement centers, and from your friends. Unfortunately, because each cover letter is context-specific, it is hard for you to judge just how well the writer has hit the target.

Consider the example job ad and cover letter below. A version very similar to it was featured in a packet of example materials sent out by a respected career planning and placement center. You might think that such a letter would be perfect. You would be WRONG! It is far from abysmal, but it has a few mistakes, flaws, and inadequacies that reduce its effectiveness.

Job search nightmares.

Exploration Geophysicist-Level IIIA

Slumberdeeply, Inc.

Position is an entry-level technical position in the Borehole Technologies group of Slumberdeeply, Inc. Candidates must have strong technical training at the M.S. level with emphasis on mathematics, mechanical engineering, geophysics, and computer science. Specific required skills include programming ability and experience in C and Pascal, manipulation of large data sets, and knowledge of exploration geophysics technologies. Desired skills include experience in a product engineering environment, excellent communication and teamwork, and a desire to join a fast-paced technology company.

November 5, 1999

Human Resources Department
Slumberdeeply, Inc.
600 Snoozy Lane
Inferno, Texas 54321

Dear Sir or Madam,

I am interested in learning more about technical job opportunities at Slumberdeeply, Inc. I believe my coursework, interest in geology and computers, and previous experience would be of value to your organization. I am especially interested in the Borehole Technology Engineer position, which was recently posted at Mighty University's Career Planning and Placement Center.

I have recently completed a Masters of Science degree in Geophysics from Mighty University. My area of specialized courses included computational seismology, geodesy, and remote sensing. Beyond the geology and physics courses I have taken as an undergraduate at MIT, I have learned programming languages (C, Pascal, FORTRAN) and operating systems (UNIX, DOS, Macintosh). Through my involvement in my research group's computer network, I have been able to keep up to date about the latest software, including Microsoft Word and Lotus. I therefore believe that I would be very effective in the position that was described in your job announcement, which called for experience in these areas.

Enclosed please find my resume for your consideration. I would appreciate the opportunity to discuss my qualifications with you. I will call your office to follow up on this letter and to explore how I can meet your needs. If you have any questions I can be reached at (408) 555-4321.

Thank you for your consideration.

Sincerely,

Geophysics Graduate Student

The letter above suffers from a number of specific problems:

1. The letter is addressed to the HR department, not the technical supervisor who is filling the position (the writer doesn't even know the sex of the individual who will read the letter!).

2. The letter does not discuss specifics. The student describes coursework and some computer experience but no specific projects or training that fits the needs of the employer.

3. The writing is wordy and too general. Sentences such as "I therefore believe that I would be very effective in the position that was described in your job announcement, which called for experience ..." is verbose and not specific. Plus, it doesn't exactly demonstrate excellent communication skills!

4. The letter contains clichés, such as "Enclosed please find ...". Such clichés are unnecessary.

Now consider an alternative scenario. Our geophysics graduate student had read the fine book that you are holding. She would

- Know that it is critical to find out as much about the job opening and the company as possible *before* submitting a resume and cover letter.

- Have called the contact person on the job ad and asked who was the chief scientist or engineer supervising the job opening.

- Have talked to the geophysics professors in her department and asked whether any of them knew the chief scientist or had information about the research group.

- Have gone to the career planning and placement service and looked up information on the company

- Have checked out their Web site.

- Have checked the career center to see if there were any former students who were working for the company.

THEN she would have called the chief scientist, asked him specifics about the job opening, such as qualifications, whether they would consider someone with engineering experience but not an engineering degree, and so on. Then and only then would she have sat down at the computer and written her cover letter. It might have looked something like the following:

November 5, 1999

Dr. Louis Cypher
Slumberdeeply, Inc.
600 Snoozy Lane
Inferno, Texas 54321

Dear Dr. Cypher:

Thank you for speaking with me last week about job opportunities in the Borehole Technologies group at Slumberdeeply, Inc. The Borehole Radar project, for which you currently have an entry-level opening, is of great interest to me. As you requested I am sending you my resume. I believe my extensive engineering training, research experience, and interest in the field of borehole geophysics would make me an asset to your group.

As my resume indicates, I recently completed an extensive Master's of Science program in Geophysics at Mighty University. In addition to coursework in computational seismology, numerical analysis, and mechanical engineering, I carried out independent research studying the seismic structure of the Great Basin. This work required me to master several computer languages, including C and Pascal, and develop reliable interface and processing software. In addition, as assistant system administrator in the Department of Geophysics, I had hands-on experience managing a diverse network of 25 UNIX workstations and mainframe computers. These skills directly address your needs for an engineer with extensive computer experience.

In addition to my coursework, I have additional experience as a product engineer. During the summers of my junior and senior year at MU, I worked for Sugardaddy Electronics assembling, testing, and troubleshooting diagnostic equipment used in manufacturing. This job not only gave me work experience in the demanding environment of commercial engineering, but also taught me the value of team-oriented problem-solving.

I am excited at the prospect of working for Slumberdeeply, Inc. I think my outstanding technical training and my enthusiasm for the field of borehole geophysics would be an excellent match with your needs. I will call you in 2 weeks to discuss what possibilities might exist for me at Slumberdeeply, Inc.

Thank you again for alerting me to the job opportunities at Slumberdeeply, Inc.

Sincerely,

Geophysics Graduate Student

Conclusion

The cover letter is an opportunity to take control of your message. It is a chance to guide the potential employer to your strengths, to answer their questions, and to look your best. Your first cover letter may take a long time and require several drafts to get right. However, once you have produced a few, the process of writing an effective cover letter will come much more easily. Just remember: without information about the job and the company, you're shooting in the dark.

Summary

- Your cover letter should provoke the interest and curiosity of the readers and guide them to your greatest strengths.
- Information about the job and the organization is essential to writing an effective cover letter.
- Like your other job materials, your cover letter must look professional and dispel any negative impressions the readers may have about scientists

Relax.
AJL

The Interview and Beyond

12

There are two terrible places to be during an interview—sitting in front of the desk wondering what on earth is going to happen next, and sitting behind the desk asking the questions. The average interviewer dreads the meeting almost as much as the interviewee ...

Martin Yate
Knock 'Em Dead

So you got an interview! Congratulations! Far from panicking, you should pat yourself on the back for a job well done. Getting an interview is an indication that you are at or near the top of the list of candidates for the job for which you applied. It also may be the first official "validation" of your credentials that you have received in your search.

Getting an interview may also fill you with dread! Up to now, your job search has been on your terms—now someone else will control the agenda. Scientists applying to non-science jobs may face additional anxiety. Your technical qualifications and your brainy background may have elicited the curiosity of the employer, but it is in the interview where employers will try to determine if you are something more than just a scientist.

Be Relaxed, Be Confident

If you feel awkward in unfamiliar social situations or get tongue-tied when speaking to strangers, you may worry that you will be at a significant disadvantage compared to other applicants. You may also feel that, while they did call you for an interview, it will be the slick, glib, and smooth candidate who gets the job. Not necessarily. Realize that the employer is looking for the best person for the position, not a used car salesman. Maturity and confidence are far more important than a winning smile and the right necktie. Realize also that the people who interview you may be as uncomfortable in unfamiliar situations as you are. The only advantage they have is that they are sitting on the other side of the desk.

The goal of an interview is to get you a job offer or at least another interview. If you handle the interview well you will show the employer that, in addition to your outstanding technical qualifications and background, you are a good communicator, an organized, prepared, and logical thinker, and someone who would add value to the organization. Some technically trained people worry that they need to "fit in" to an organization to succeed. In some ways this is true: employers are looking for people who will work well with the other members of the team, enjoy the environment, and want to stay. But fitting in does not necessarily mean becoming an anonymous cog in a machine! Many companies value individuality and initiative when it helps the team. Don't feel that you need to show an interviewer that you can "talk the talk and walk the walk." Try to show them how your unusual background and training will be an asset.

Be Prepared

In interviews, as with the rest of the job hunting process, careful research is the single key to success. Research does not simply constitute knowing the vital facts of the organization; it involves understanding the corporate culture, the work atmosphere, the mission of the organization, and doing your best to show how you would add to it. Careful research may enable you to anticipate some questions in advance. No amount of the gift-of-gab can substitute for this sort of preparation.

If you are interviewing with a company you have already researched extensively, then you probably know a great deal about the organization. You may have gone on an informational interview and may be on familiar terms with some of the people who work there. If so, you are at a great advantage—you are aware of the culture of the organization and their goals. Perhaps you know their reason for calling you.

If, on the other hand, the call for an interview came from a more chance encounter, say, from submitting a resume without any prior information about the company, then you have your work cut out for you. You must learn more about them before they see you.

As you prepare for an interview you will realize just how important informational interviewing can be. As explained in Chapter 6, informational interviewing gives you the opportunity to explore the organization on your terms. Only in this way can you learn the subtle parts of an organization's culture; for example, what the typical attire is in the workplace. Informational interviews are also good practice for the real thing. Furthermore, if you did an informational interview, then submitted your resume, then were called for an interview, you would have some indication that you made a good impression. This should build your confidence.

The Ethology of Job Interviews

Most scientists are, at best, ambivalent about appearance. Academia certainly doesn't teach us many good lessons in the fashion department. One professor in my department wore purple hiking socks and sandals every day of his life, even at scientific meetings! And the hairstyle! Oh, my!

Although you may disdain fancy clothing, you have to admit that, for most other animals in nature, appearance is a life-or-death issue. Animals judge each other's health, strength, and parental fitness not by subjecting their potential mates to an interview but by examining certain external qualities that are a proxy for fitness. In the animal world, these qualities include thick, well-groomed fur, long feathers, big antlers, or, in the case of the stickleback fish, the size and hue of the red spot on his belly.

People evaluate each other all the time using similar traits. Even the most thorough job interview does not give an employer enough information to reliably test an applicant's total abilities and potential. Choosing whom to hire ends up being a very subjective process, based on a whole range of judgments about externalities that may or may not correlate to ability. The TV news program *60 Minutes* conducted an undercover investigation of this phenomenon and showed that good-looking, well-groomed candidates were offered jobs three times as often as similarly qualified candidates who were less polished in appearance.

The beauty of being human is that, unlike much of the rest of the animal kingdom, we are endowed with opposable thumbs, large brains, and the ability to purchase on credit. As a result, you have the power to dramatically change your appearance, both to suit the situation and influence how others perceive you.

Some young scientists might feel weird spending the equivalent of one month's salary on a suit, or going to a salon for a full makeover. Get over it! You may be smart, but you should also use all the weapons at your disposal to positively influence prospective employers. This includes clothing, personal appearance, manners, and body language.

Communicating with More than Your Mouth

As with fashion, many young scientists are equally unfamiliar with aspects of body language and how they affect others' perceptions. While nobody enjoys feeling totally self-conscious, everyone should be aware of their own body language and what it says to others. Here are a few issues that seem to be a problem for some young scientists:

- *Eye contact*
  Solid eye contact conveys the message that you consider yourself, and the person you are speaking with, important. Looking people in the eye when they are speaking makes them feel "heard." Looking at them when you speak lets the person know that you care that they understand what you are saying. You needn't keep constant eye contact or mutter "one-Mississippi, two-Mississippi"

under your breath but you should try to keep a focus on the other person. Here's one good strategy to use at the beginning of an interview: when you shake the interviewer's hand, make a mental note of the color of his or her eyes.

Lack of eye contact can communicate a variety of negative messages. Many will assume you lack interest in them or confidence in yourself. Sometimes you can look dishonest! Foreign-born scientists take note: in other cultures, looking directly at and maintaining eye contact with someone who is a superior can seem presumptuous and rude. But in the United States, a deferential aversion of your eyes will probably be misinterpreted. Be confident, and, if you are having trouble being comfortable maintaining eye contact, seek help through practice interviews.

- *Focus*
  This is more than eye contact. Focus is maintaining your attention on the interviewer and the questions he or she is asking. Focusing can be very hard during an interview! You are nervous, you are wondering about your appearance, how the interviewer is reacting to you, and, of course, whether or not you'll get an offer. Under such circumstances it is easy to appear distracted (you are!) and to lose the train of conversation. There is no easy remedy for this other than lots and lots of practice! Being comfortable with your message and familiar with the organization can help reduce your anxiety.

- *Posture*
  "Sit up straight!" I heard this a lot (I am a big sloucher). Like eye contact, good posture (squared shoulders, direct, and erect but not rigid or tense) suggests that you are both alert to the interviewer and that you have something to contribute. When sitting, you should avoid a very restricted, symmetrical position with feet together straight ahead and hands clasped in your lap.

- *Gestures*
  Like your posture, your gestures should be smooth and confident, not spastic and excessive! Beware the bouncing foot! Interviewers can see that nervous 4-Hz foot bounce even under a table. It is often impossible to be fully aware of your true gestures, especially in an interview situation. That's why a good practice interview is so valuable. Many career planning and placement centers can videotape your practice interview so that you can see exactly how you look in the hot seat.

What Should I Wear?

There is so much advice about the professional wardrobe. There are books, consultants, and information available on-line. *The New Professional Image*, by Susan Bixler, is an excellent guide to professional attire for men and women. I have also found that knowledgeable sales assistants at the higher-end department stores are wonderful sources of advice about business attire; after all, they know what's popular.

But knowing what to wear often comes down to knowing what's appropriate. A business suit is probably a good bet for a man interviewing at an investment bank, but what about for one at a biotech company? If you have conducted some research on the organization, gone on an informational interview, and checked with your network, you will likely have some idea of what the general fashion expectations are.

Attire for an interview is more formal than what's worn for a typical day on the job (no surprise there). But your attire cannot be 17 rungs above the typical! An informational interview can help you gauge where the "everyday" fashion of the organization is. Taking the right fashion step above that may be obvious or subtle, so it is valuable to discuss the issue with friends and consult the sources identified above.

What Kind of Interview Is It?

All interviews are not created equal. You may be asked different questions depending on where you are in the selection process. Interviewers usually are willing to tell you ahead of time what kind of interview to expect. Knowing in advance can help you to go in calm and confident. Here is a description of the most common types of interviews:

- *Screening interview*
  If you are participating in on-campus interviews, or if you are one of many people applying for a job, the first interview may be to cull the herd. Screening interviews, as they are known in the parlance of recruiters, are brief (usually 20-30 minutes) and designed to be a first-pass through the applicant pool. Typically, employers have a few specific questions they want to ask. For scientists with an advanced degree, the most obvious question might be "why are you applying for this job?"

- *One-on-one interviews*
  The one-on-one interview is the most common type of interview, involving just you and a single person asking questions.

- *Phone interview*
  In some cases (usually a question of time and cost) an employer might want to talk to you first over the phone. This does give you the advantage of sitting in your own environment and wearing anything that you want (how will they know you are dressed up like Xena the Warrior Princess?) But the same rules regarding formality of questions and answers in regular interviews apply in phone interviews. Phone interviews should be scheduled in advance, like any interview, to allow you time to prepare. You should conduct them in a quiet place where there will be absolutely NO DISTRACTIONS.

- *Panel/committee interviews*
  This type of interview, involving you and several interviewers, is less common for entry-level jobs than for higher levels of employment. They can be a bit more stressful, if for no other reason than

you are outnumbered (sort of reminds you of your qualifying exams in grad school, doesn't it?) However, they can also be more enjoyable because of the variety of people you have in front of you.

- *Case study interviews*
  In some fields, such as finance and consulting, the case study interview is the norm. Rather than ask you about yourself and your background, the case study interview presents you with a situation, usually a typical business problem or dilemma, and then asks you to provide some logical structure for solving the problem. The objective is to observe your approach to problem solving (preferably logical and organized), to evaluate your analytical abilities (some may provide calculators and paper), and to see how you arrive at a logical conclusion. It is rare that you will have any specific knowledge about the industries or subjects you are asked about, so it is important to ask salient questions, follow a logical process of evaluation, and by all means, arrive at a conclusion. Some career planning and placement centers run workshops on how to prepare for such scenarios.

- *Technical interviews*
  In many programming jobs and some other technical areas some or all of the interview may involve solving specific technical problems, such as writing code or using a specific software application. It's always nice to know you're going into this sort of situation beforehand...

- *Stress interviews*
  When is an interview not stressful? You may have a different answer to that question after surviving a stress interview. Stress interviews are designed to see how you stand up to pressure. They may involve difficult questions, or an impatient interviewer, or an interviewer who deliberately tries to destabilize you. Typically, you may encounter a stress interview if it is one of several interviews for a specific job. It is very hard to prepare for such an encounter, but at least knowing that stress interviews can be expected from time to time can be some comfort. If you realize that someone is trying to stress you out, just take a deep breath, realize that it is their aim to see you under pressure, and be COOOOOL.

Many job interviews are actually a collection of meetings with various people in the organization. In academic interviews you might have hour-long discussions with nearly a dozen people over the course of 2 days. Interviews for research positions in industry and national labs may have a comparable schedule. In addition to these, you may be invited out to lunch or dinner with a group of people. Even though these encounters seem more social and less formal, you ARE being scrutinized! So, save that off-color joke or those disparaging remarks about the waiter until AFTER you are hired!

What Will They Ask?

As stated earlier, if you've done your homework, you may have some idea about what they will ask you. If you have had other interviews you may already have some fairly good responses down pat. For the scientist applying for a "non-traditional" job that does not require an advanced degree, or for that matter, a science degree of any kind, the first and most obvious question might be, "why do you want to work here?" "Because I need a job" is not the best response. No matter where you go to interview you should be prepared to answer this question. As with any response, the aim is to showcase your abilities.

Knock 'em Dead, by Martin Yate, is one of the most popular books in print about interviewing. Almost half the book consists of "Great Answers to Tough Questions." I was surprised how often variations on one of Yate's questions came up in my own experiences, and other friends have told me the same thing. This book is a good one to check out of the library to help you prepare. Some of the most common questions asked of scientists interviewing for non-science jobs are below, along with some suggestions about how to answer them:

Aren't you overqualified?

- Explain how you are HIGHLY qualified, but not overqualified.
- Tell them that you will be up to speed quicker.

What is your greatest strength?

- Talk about one of your best skills that RELATES to the job.
- Give a concrete example of how you used it.

What interests you about this position?

- This one seems straightforward, but you'd better be sure you KNOW enough about the job.

Describe a situation in which your work or idea was criticized.

- Choose an example that involves constructive criticism of your work.
- Be sure to tell them how you rectified the situation.

Describe a project in which you demonstrated _______ . (Fill in the blank: leadership, teamwork, initiative, problem-solving skills, ability to take criticism, etc.)

- Tell a story (a SHORT one) about a specific incident, not just about what you did, but about the final result.

What is your greatest weakness?

- Save absolute candor for your therapist. Describe a weakness that could be considered a strength.
- Avoid cliché answers like, "I'm a perfectionist."
- Show how you have compensated for the weakness, perhaps by relying on help from others in the team.

Why are you leaving research science?

- Accentuate the positive; don't tell them you were miserable, or that you could never find a permanent job. Tell them that you are looking for new challenges and a place where you can apply some of the knowledge you learned in grad school.

So tell me about yourself.

- Focus on how your background relates to the job.
- Limit your answer to less than 3 minutes.
- Try to explain how your background and experience has directed you to this job.

Practice Makes Perfect

Would you give an important talk at a national or international meeting without practicing? Of course not! Then why would you go off to a job interview without running through some of these questions with a friend? Practice works, especially if you have an audience that can give you feedback. You can practice answering interview questions either with friends or a career counselor. Some career centers even let you videotape yourself! This may sound really embarrassing, but it is actually both fun and immensely valuable.

Techniques for Answering Questions

Remember that your goal is to use specific examples to back up the claims you made in your cover letter and resume. One good way of approaching responses to questions such as, "Describe a situation or project that showed your ability to ___," is to structure your answer. A good structure to follow is called the **STAR** approach:

Situation/**T**ask: Describe the situation you encountered. Give the background and its relation to you.
Action: Describe what YOU did to address the situation or solve the problem.
Result: Describe the result of your actions.

Structured in this way, you give the interviewer a short, structured story that will only take a minute or two.

Telling a story in your answers is an extremely effective communication tool. People like stories. Stories have a beginning, middle, and end and can contain dramatic tension and even humor. Telling a story puts you, the storyteller, in control of the situation. Humor is a valuable ingredient. Being able to laugh at yourself and being comfortable enough with yourself to admit your past mistakes is an indicator of maturity and confidence. Just don't tell them a story that's too embarrassing!

An Example

Bill Dos (see page 116) was interviewing for a job with an oil company that used gravity data sets for resource exploration. At one point, the interviewer asked Bill:

Tell me about a situation in which you showed initiative.

Bill responded:

> *One example is my development of the PointerCalc software package, which I did while in graduate school. When I was working on my thesis research I had to develop several computer programs to manipulate*

and project the gravity data I was using onto a map or globe. After finishing up the programs, other people began asking me for copies so that they could do the same manipulations with their data. While I could have simply given them copies of my programs, I realized that some of their data sets would not be compatible, so instead I rewrote my programs to accommodate a wider variety of vector data. I did this mostly because I enjoyed programming and because I thought it would be so much easier if one person made a general program rather than everybody having to make specific variations in my code to suit their individual needs. Plus, it didn't take very long. After circulating my revised software throughout the department, I began getting requests from researchers in other institutions! After one year, I had over 1000 registered users on the software in a variety of fields. While I have always provided the software and its upgrades at no charge, I have ended up benefiting enormously from their popularity.

What Shouldn't They Ask?

There are a number of laws aimed at preventing employers from unfairly discriminating against applicants. Some of these laws proscribe certain questions from being asked in an interview situation. However, illegal questions pop up surprisingly often, most often unintentionally. According to the law, an employer may not ask you questions about:

- Your religion, political beliefs, or affiliations
- Your ancestry, national origin, or parentage
- The naturalization status of your parents, spouse, or children (they can ask whether or not you are a U.S. citizen or the status of your visa)
- Your birthplace
- Your native language (they can ask about the languages you claim to speak on your resume)
- Your age, date of birth, or the ages of your children (they can ask whether or not you are over 18)
- Your maiden names, or whether you changed your name, your marital status, number of children, or spouse's occupation (this is the most commonly encountered illegal question asked of female job applicants)

If you feel that you are being asked a question that crosses the line it is not necessary to blow your police whistle and read the person the riot act. Most often, inappropriate questions are asked because the interviewer is inexperienced or clueless. Jumping down their throats may both offend them (after all, they didn't think they were doing anything wrong) and stigmatize the rest of the interview.

Inappropriate questions often arise because of some underlying concern on the part of the interviewer. For example, an interviewer who (inappropriately) asks "are you planning to have children?" may not care so much about your childbearing capabilities but might be more concerned about how the demands of the job will affect someone with a family. Rather than directly confronting this question you might answer by trying to allay such underlying concerns. For example, you might respond:

> *If you are concerned that I may not be able to work overtime on short notice, let me assure you this is not a problem.*

If you are genuinely unclear as to where the inappropriate question is coming from you might ask:

> *I'm not sure of the relevance of that question; can you tell me how it specifically relates to the job?*

Or you can legitimately and politely refuse to answer the question.

A few years back a young scientist in one of my workshops raised her hand and offered her own clever solution to such inappropriate questions which she had often been asked when she was interviewing for her postdoc:

> *The interviewer asked me if I planned on having children. I looked him right in the eye, smiled, and replied: "You sound just like my mother-in-law!" He laughed, I laughed, and the issue did not come up again.*

What Should I Ask?

Just like the informational interview, the job interview can be a two-way flow of information. Having some questions ready is a sign that you are prepared and that you are interested in the position. Most of the questions listed in Chapter 6 are just as good here. In addition, you should probably find out what the next step in the process will be. Will they call you? If so, when?

Send a Thank You Note After the Interview

Send a short thank you letter to the person or people you talked to unless specifically asked not to do so. Do it within 3 days of the interview or don't bother. Some people couldn't care less, but a few care a great deal. Your thank you letter should be short, and should, in addition to thanking them for their time, highlight your qualifications. This act of courtesy shows attention to detail and will set you apart from the rest of the applicants.

Some Final Advice on Interviewing

- Arrive early. Give yourself 10-15 minutes to sit and chill out.
- Case the joint. If it is in a place you've never been before, swing by the day before just to make sure you know how to get there.

The assurance of having been there before will help.

- Bring along extra copies of your resume.

- Give a good handshake. If you are unclear about what this is, go try out your handshake on your friends.

- Make eye contact. One simple technique for ensuring that you have made good eye contact: make a mental note of the color of your interviewer's eyes.

- Ask questions. It's better to be clear about the question at the start than to ramble off on some tangent.

- Be yourself. People tend to do a poor imitation of anything else.

The Other Side of the Interview

While job interviews ARE under the control of the employer, information and evaluation is a two-way process. The savvy candidate is not only being interviewed but is INTERVIEWING the company. Like the interviewer, you, the interviewee, should be on the watch for telltale signs that the job and the organization may not be such a good opportunity after all. There's only one thing worse than losing out on a great job opportunity, and that's accepting a job that turns out to be nightmare!

How the organization handles their interviews speaks volumes about how they value people. Some organizations realize that the BEST candidates will likely have multiple offers. If they want to hire the BEST, they have to make a strong sales pitch. Smart companies try to make a good impression on ALL job candidates. They know the value of people and they tend to have a workplace and a workforce that reflect that.

Many other companies, however, fail to realize the importance of making a good impression, or, they may simply blow it in the execution. Some organizations believe that they are SO GOOD, or that the job market is SO BAD, that they don't even have to try to make a good impression (academia is notorious for this!).

Most job seekers are so happy to get an interview that they will put up with practically any behavior on the part of an interviewer or an organization. They fail to realize that the values and capabilities of the organization are reflected in the interview process. The interview is a window into the soul of an organization.

Being a Freudian Job Seeker

Like any good psychoanalyst, you should go into an interview looking for clues to the underlying psyche of the organization. Here are some questions to consider:

- How well is the interview process organized?
- How do they behave?
- What do they tell you about themselves?
- What do they ask you?

Some companies run a very tight ship.They notify finalists when they say they will. Recruiting information is sent in advance of the interview. The day of the interview, candidates are given an itinerary, and all the people on the list have been fully briefed about them. Afterwards, candidates are told when to expect results and those results arrive on time. Such an organization clearly values people and their time. They are efficient, organized, and—most importantly—allocate the resources to do the job right.

Consider the alternative. The organization takes months to figure out the short list. Candidates are given no background material on the people they will meet, and the people who interview them seem unprepared and distracted. After the interview, the organization delays in making the final decision, leaving the finalists hanging (that is, those who have not already chosen a better opportunity). Such an organization places a lower value on people or may not have the resources available to do a better job. If they're so poor that they can't even run an interview right, I wonder what your prospects would be for getting a new initiative funded.

People's behavior during the interview process also speaks volumes about the values of the organization. Some places present a friendly, positive face, and all who interview candidates make a point of mentioning the values of the company. Other organizations present a tough, confrontational side. The worst organizations are careless and unprofessional. They might leave an interviewee sitting in a room for 45 minutes because someone missed an interview appointment. What do you think that says about how the operation is run?

Finally, an observant person can learn volumes by listening to what interviewers say about the organization. Do all the interviewers have a consistent message? Are all of them happy with their work? One friend of mine sat through a 45-minute interview in which the interviewer complained incessantly about not being promoted! Another friend heard from three of the five interviewers she spoke with that the organization was so chaotic that they literally did not know what would happen next. Hmm, sounds like a GREAT place to work!

People will also reveal interesting inconsistencies between the external face of the organization and its true internal psyche. For example, the organization may claim to value research and development, but it can become clear in your conversations with people where the true priorities are. Typically, at the end of an interview, the interviewer will ask you if you have any questions. Ask them, "what is the best thing about working in their organization?" and watch the reaction. Are they surprised by the question? Do they have trouble answering it? If they do answer it, is the answer something that you think is an asset?

Finally, the questions they ask of you reveal a great deal not only about what they think of you but also how they think about employees in general. Some organizations emphasize teamwork and a cooperative environment. Some even administer a self-assessment instrument (see Chapter 5) to determine your general personality type. Other organizations ask only technical questions. In those cases you should worry whether they CARE about other aspects of the job besides technical skills.

If the interview process really leaves you feeling weird about the organization, perhaps your subconscious is telling you something. Perhaps the job description is fine but the environment just doesn't seem right. Listen to your subconscious! No matter how desperate you are, there will be OTHER opportunities. A job, like a marriage, should not be entered into lightly.

Negotiating an Offer

"Well, Ms. B____, we were very impressed with you and we'd like to offer you the job." PSYCH! PARTY! WAY TO GO! You've done it. But before you hang up the phone and pop open a beer, you may want to consider THE OFFER.

For young under- or unemployed scientists, an offer, ANY OFFER might seem a true blessing. It may be so, but a job offer can be negotiated. Typically, entry-level positions have a fixed salary and benefits package attached to them. However, you'd be surprised how much wiggle room you may have in negotiating. It is also important to remember that compensation includes a number of things besides salary. Often employers are most stingy with salary but can be surprisingly generous with other benefits such as paid vacation, health care, equity participation, and other incentives.

Nobody is forcing you to haggle for the best offer possible. But realize that, once they have made you an offer, your value to them has grown tremendously. Not only is it important to capitalize on that, but salary negotiation itself is an important part of professionalism in any career. It is highly unlikely that a polite inquiry on your part is going to send the potential employer off in a huff.

Delay Salary Negotiations as Long as Possible

Your value to the employer rises with each stage of the interview process that you pass. Employers often try to lock you in to a certain level of compensation before they actually give you an offer. It is best to avoid this by simply stating that you would prefer to discuss salary and benefits once a firm offer has been tendered.

Value Yourself and the Job Properly

The job for which you have just received an offer probably has some salary range attached to it. This range is to compensate for skill and experience. Though you may not have any direct experience doing what they want you to do, you do have lots of other experience. Some of this may "count" in their consideration of a salary for you, and it is important for you to point this out. There are also a large number of other aspects of compensation that should be considered such as:

- Health care: who is covered, what's covered, what's not, and what are the co-pays and premiums?
- Schedule of raises
- Bonus plan
- Commission plan
- Stock option
- Pension plan
- Profit-sharing plan
- Employee education/tuition reimbursement
- Dependent tuition reimbursement
- Paid parking
- Car provided
- Vacation
- Sick leave
- Maternity/paternity leave
- Flex time/alternative work schedule
- Anticipated work hours
- Relocation allowance
- Potential for advancement
- Stability of company

You should ask questions about all these issues *before* you settle on a salary.

Finally, if you are entering a career in which you really don't have a good idea what the salary range is, find out! There are a number of guides and surveys available on the Web and in career centers (for example at Science's Next Wave-www.nextwave.org). Your network is also a good source for timely salary information.

How to Get the Offer Raised

The employer is trying to "buy" you at as low a price as possible. You should be trying to "sell" yourself at as high a price as possible. Given the list above, you should take a hard look at the total offer. Is there a reason why you might accept a lower salary in exchange for something else, such as a yearly education stipend? On the other hand, is the salary undervaluing some aspect of the job, such as long hours or lots of travel?

Next, consider the factors listed below. The more that are true, the greater your flexibility:

- You possess unique abilities.
- They have few other candidates for the job.
- The search has been going on a long time.
- This is a unique position in the organization.
- The organization is flexible in general.
- You have other offers.
- They really need someone soon.

In contrast, you will have less flexibility to negotiate salary and benefits if the following are true:

- The job is at an entry level and similar to others in the organization.
- The organization is highly structured and rigid.
- The organization expects you will take what is offered.

Conclusion

Remember that you have spent a number of years obtaining a very valuable education and have carried out complex projects successfully with little supervision. You are different from the other applicants they are considering and more valuable because of it. You are the best and the brightest that this country has to offer. Don't let them forget it!

Summary

- Do your homework: know where the interview is, what kind of interview it will be, and what the expectations are.
- Practice answering some commonly asked questions ahead of time.
- Don't be afraid to negotiate, especially for things besides a higher salary.

Smith
Investment
Banking
Smith
Research
Labs

13 Perceptions and Realities

Concern for man himself and his fate must always form the chief interest of all technical endeavors ... in order that the creations of our mind shall be a blessing and not a curse to mankind. Never forget this in the midst of your diagrams and equations.

Albert Einstein
Address
California Institute of Technology (1931)

Many people have a conflicted view of those of us with a Masters or Ph.D. in science. We, who have lived and worked with other advanced degree holders, tend to take the M.S. or Ph.D. somewhat for granted. However, most people have limited experience with advanced degrees and the people who have obtained them. Most people assume that in order to obtain an M.S. or Ph.D. you have to be VERY smart. Some people can be downright intimidated. While we know that people with advanced degrees have a wide range of intelligence and ability, the average citizen assumes that we are all rocket scientists. This presumed brilliance is not such a bad thing; it can work to your advantage in your job search.

On the other hand, people also hold some rather negative stereotypes about those of us with advanced degrees. Most people believe scientists are nerdy and out of touch with reality. Just look at the movies, or TV, and you will see how our culture views us: smart, odd, funny-looking, awkward, and overly focused. Some people have a darker view of science in general and scientists in particular. Popular culture in Europe and America has been full of images of the mad scientist bent on world conquest or world destruction. Nearly 100 years ago, Marie Curie was lampooned and somewhat demonized in the French newspapers of her day for her work on the mysterious phenomenon of radioactivity. And negative stereotypes of scientists persist today.

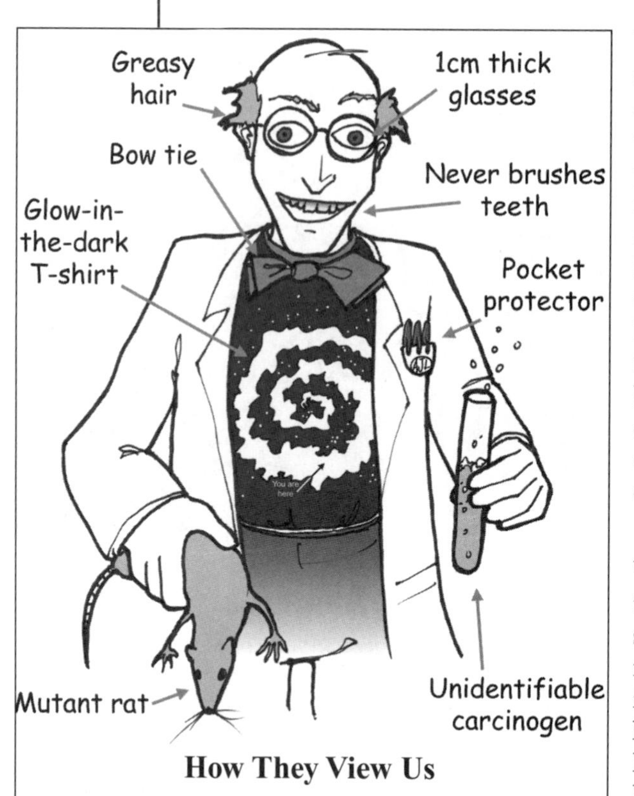

How They View Us

When an average person thinks of a scientist, the image that he or she conjures may look somewhat like the cartoon at the left. Not very pretty, is it? This stereotypical image has all the derogatory features that people associate with us: funny clothing, pocket protector, ugly glasses, funny haircut. However, there is another stereotype about us that isn't necessarily bad: everyone assumes that we are extremely smart. This is the essence of the conflicted image our fellow citizens have of us.

When people unfamiliar with scientists and the scientific career first see the Ph.D. or M.S. on your resume, an image not unlike that featured above might pop into their heads. However, if your resume looks professional, you will challenge their preconceptions that you are a clueless geek. When they talk to you over the phone and you sound NORMAL, you will further undermine their negative preconceptions. But when you walk in the door for your interview looking professional and sounding articulate and confident, you will banish their negative stereotypes forever. All that will remain will be their idea of you as very, very smart.

Try not to dispel that preconception if you can.

Hiring managers often have other more subtle concerns about advanced degree holders applying for "non-traditional" jobs. Some may fear that you are a hermit and will not interact well with others. Some may worry that your years in the ivory tower have made you arrogant and incapable of working on a team. They may fear that you are disorganized, simple-minded about money, and incapable of coping with deadlines. In your resume, and in all your subsequent encounters with people unfamiliar with a scientific career, it is important to demonstrate how you do not possess any of those qualities they fear.

Challenging Your Own Stereotypes

Scientists are every bit as guilty of harboring negative stereotypes as anyone else. Academicians are notoriously guilty of looking down on all other career fields. Mostly, that opinion stems from ignorance. For example, you, or some of your mentors, may think that a career in business would require you to be slick, superficial, money-driven, and smarmy. In reality, people in business come in a wide assortment of shapes and sizes. Some of them are smart, too! You may think that anyone who has left graduate school without finishing his or her degree has somehow "flunked out" no matter what the circumstances. These ill-informed, unsubstantiated opinions end up hurting you more than anyone else, by limiting your options as well as reducing the number of friends you have!

How We View Them

You Are More than a Scientist

In my travels I have met hundreds of M.S.- and Ph.D.-trained scientists who now have fulfilling and challenging careers in a huge array of professions. They are high school teachers, stock analysts, government officials, journalists, members of Congress, entrepreneurs, writers, lawyers, and more. I even know a rodeo star with a Ph.D! Many of them have found that the skills they acquired as part of their training in science have wide application beyond science. More importantly, most have found ways of contributing to their communities, and to society, by applying their skills outside the world of science.

We will always need brilliant, creative, and energetic men and women to carry scientific research forward. However, we will also need brilliant, creative, energetic people to carry the lessons of science to the rest of the world. With your training and experience, you can have an exciting and fulfilling career doing either. Or both. Don't be afraid to consider a "non-traditional" career just because it is unfamiliar to you, your advisor, or your family. There are so many important problems and opportunities that are calling for the attention of smart people who want to make a difference.

The Scientist in the "Real World"

For those of you who have already made the transition to a career outside science, as well as those who manage to do so after reading this book, realize the important role you might play in guiding the careers of other scientists who are exploring their career options. Making a major career shift is difficult and painful, even with an abundance of career advice from books and counselors. The most potent and valuable guidance often comes from those who have actually lived through it. Here are some ways in which you can help:

- Sign up as an alumni contact with the school at which you received your advanced degree or degrees. Until now, nearly all the career services and contacts provided by colleges and universities have been geared toward undergraduates. Graduate students need the same services and support.

- Remain a member of at least one scientific society. Scientific societies tend to dwell on the needs and concerns of the research community because that is where the bulk of their membership lies. However, a small and vocal group of "outsiders" would not only help to broaden the scope of the organization, but they would serve as important role models for younger members considering alternative careers.

- Hire other scientists. You, more than anyone else in your organization, can appreciate the value as well as the difficulties of hiring and training scientists for non-science careers. If the "old boy/old girl" network can work for graduates of the Ivy League, why can't it work for the scientific community?

- Don't be a "former scientist." While you may not be carrying out basic research and publishing in scholarly journals, the basic skills you developed as a scientist will remain with you forever. Use them and show others how to do the same!

Summary

- In the "outside world" people may have many preconceptions of you because you have an advanced degree in science. It is critical to dispel the negative stereotypes by presenting yourself as professionally as you can.
- You probably harbor your own stereotypes about other career fields and the people in them. You should challenge your own stereotypes as well.
- Once you have moved to a non-research career, you could help others tremendously by being a mentor and role model. Stay connected. There is no such thing as a "former scientist."

References and Resources

Science Policy, Funding, and Job Market Trends in Science

Atkinson, R., Supply and demand for scientists and engineers: A national crisis in the making, *Science, v. 248*, 1990, pp. 425-432.

This article is cited by many as a prime example of how "The Myth" of massive retirements and numerous academic job openings was promulgated. Too bad the data was flawed and the analysis was wrong.

Browne, M. W., Supply exceeds demand for Ph.D.s in many science fields, *New York Times*, July 4, 1995, p. 22(N).

One of a number of articles from the mid 1990s that highlighted the glut of Ph.D.s.

Carnavale, A. P., Gainer, L. J., and Meltzer, A. S., *Workplace Basics—The Essential Skills Employers Want*, Jossey-Bass, San Francisco, California, 1990, 477 pp.

An excellent summary of job market and hiring trends in the 1990s and beyond.

Good, M. L., and Lane, N. F., Producing the finest scientists and engineers for the 21st Century, *Science*, v. 266, 1994, pp. 741-743.

Senior science policy officials in the Clinton administration, Lane and Good called for improvements in the training of graduate students in the sciences. Did academia heed their call?

Greenberg, D. S., So Many PhDs, *The Washington Post*, July 2, 1995, p. C7.

Dan Greenberg is a luminary science policy analyst and journalist. In this editorial he highlights the overproduction problem of Ph.D.s.

Horowitz, T., Young professors find life in academia isn't what it used to be, *Wall Street Journal*, February 15, 1995, p. A1.

One of many examples of how academia is changing and how those changes affect the lives of young professors.

Levin, S. G., and Stephan, P. E., Are the foreign born a source of strength for U.S. science?, *Science*, v. 285, 1999, pp. 1213-1214.

Their answer: Yes! This article shows that foreign-born scientists contribute disproportionately to science.

Massey, W. F., and Goldman, C. A., *The Production and Utilization of Science and Engineering Doctorates in the United States*, Stanford Institute for Higher Education Research, Stanford, California, 1995.

This study tried to quantify the actual demand for Ph.D.s and predicted oversupply in many fields.

Nelson, C., and Bérubé, M., Graduate education is losing its moral base, *Chronicle of Higher Education*, March 23, 1994, p. B1.

These two authors lambast the higher education system, suggesting that the production of Ph.D.s is more a function of the need for teaching and research assistants than the need for trained professionals.

Nerad, M., and Cerny, J., Postdoctoral patterns, career advancement, and problems, *Science, v. 285*, 1999, pp. 1533-1535.

Marisy Nerad has been running a fascinating study of Ph.D.s 10 years out of school. This article highlights some of her findings as they relate to young scientists.

Reshaping the Graduate Education of Scientists and Engineers. Report of the Committee on Science, Engineering, and Public Policy, National Academy of Sciences, National Academy Press, Washington, D.C., 1995.

This landmark study called for substantial changes to graduate education in the sciences. Five years later, however, academia hasn't changed at all...

Richter, B., The role of science in our society, *Physics Today*, September 1995, pp. 43-47.

The former President of the American Physical Society, Richter gives a prescient view of the possibilities open to young scientists.

Tobias, S., Chubin, D., and Aylesworth, K., *Rethinking Science as a Career*, Research Corporation, Tucson, Arizona, 1995, 157 pp.

This study examines the aftermath of "the Myth" and the state of science employment in the mid 1990s. A "must read" for anyone interested in the history of S&E employment predictions.

Job Change, Life Transitions

Bolles, R. N., *What Color Is Your Parachute?*, Ten Speed Press, Berkeley, California, 1999, 464 pp.

Parachute—as it is widely known, is a classic career development manual in the United States. Chapters 1, 2, 5, 6, and 7 all deal with coping with change. New editions come out every year.

Bridges, W., *Transitions*, Perseus Books, Reading, Massachusetts, 1980, 170 pp

Transitions is a classic that describes the difficult process of change and helps those going through it understand the process better. It may be particularely valuable for those who are finding their job anxiety spilling over into their personal lives.

Morin, W. J., and Cabrera, J. C., *Parting Company*, Second Edition, Harcourt Brace & Company, New York, 1991, 387 pp.

A good book for more senior level scientists facing job loss, job change, and similar situations.

Scott, C. D., and Jaffe, D. T., *Managing Personal Change*, Menlo Park, California, 1989, 71 pp.

These authors describe change as a process consisting of several steps. Their valuable workbook has useful exercises and is available in many career counseling centers.

Self-Assessment

Bolles, R. N., *What Color Is Your Parachute?*, Ten Speed Press, Berkeley, California, 1995 (new editions come every year), 464 pp.

Chapters 9 and 10 discuss self-assessment and contain some great exercises.

Newhouse, M., *Outside the Ivory Tower: A Guide for Academics Considering Alternative Careers*, Office of Career Services, Harvard University, Cambridge, Massachusetts, 1993, 163 pp.

This guide to career change, written by an excellent counselor with experience helping graduate students of all flavors, has great advice for scientists and non-science Ph.D.s alike and an excellent bibliography. It is available through Harvard's Office of Career Services or at most university career centers.

Keirsey, D., *Please Understand Me: Character and Temperament Types*, 6th Edition, Prometheus Nemesis, Del Mar, California, 1987, 210 pp.

This book explains the different personality types and how they interact in work and play.

Levin, A. S., Krumboltz, J. D., and Krumboltz, B. L., *Exploring Your Career Beliefs*, Counseling Psychologists Press, Palo Alto, California, 1995, 46 pp.

A guide to the Career Beliefs Inventory and how to interpret your score.

Sher, B., and Gottleib, E., *Wishcraft*, Ballantine Books, New York, 1986, 278 pp.

This very popular, somewhat touchy-feely manual on reaching your goals is a best-seller and is available everywhere.

Tieger, P. D., and Barron-Tieger, B., *Do What You Are*, Little, Brown and Company, New York, 1992, 330 pp.

This book has a chapter for each of the 16 personality types in the MBTI, describing communication style, work style, and good career matches.

Cover Letters/Letter Writing Style Guides

Krannich, R. L., and Krannich, C. R., *Dynamite Cover Letters, 3rd Edition*, Impact Publications, Manassas, Virginia, 1997, 195 pp.

Sabin, W. A., *The Gregg Reference Manual*, 7th Edition, Glencoe, Lake Forest, Illinois, 1992, 502 pp.

Interviews

Yate, M., *Knock 'Em Dead 2000*, Adams Media Corporation, Holbrook, Massachusetts, 2000, 338 pp.

Bixler, S., *The New Professional Image*, Adams Media Corporation, Holbrook, Massachussetts, 1997, 262 pp.

Perceptions of Scientists in Society

Weart, S., *The Physicist as Mad Scientist*, Physics Today, vol. 41, no. 6, 1988, pp. 28-37.

Index

D

E

F

G

H

I

T

U-V

W-X-Y-Z